"Andrea Olsen sees wide and deep, bringing us her vivid and remarkable experience of organic connection between body and earth. Written in a language both poetic and practical, this timely offering is contemporary in its unique perspective and ancient in its apparent wisdom."

—Janet Adler, author of *Arching Backward: The Mystical Initiation of a Contemporary Woman*

"Andrea Olsen takes us on an extraordinary journey through the body where the boundary between self and other begins to merge. Her book points to the deep relevance of somatics and how that connects us to the greater body in which we live . . . the earth."

—Donna Farhi, author of *Yoga, Mind, Body and Spirit*

"A simple look around will suffice to show that the human body—and the human spirit—were given without instruction manuals. Now that oversight has been corrected. This is an essential book for people who want to do more than simply exist in their bodies and in their places. It is a book about thriving."

— Bill McKibben, author of *The End of Nature*

"Andrea's writing resonates a deep engagement with the earth and our potential for connection. Wondrously matter-of-fact, she . . . gives us the tools to note those small epiphanies of science and existence, of the land and the heart, with detail, spirit and a humbling sense of place. *Body and Earth* is a practice, a reminder of the enormously simple act of conscious living. This is a book that will be rained on, danced with, left behind and found again, borrowed and used for life."

—Bebe Miller, Choreographer and Artistic Director of Bebe Miller Company and Professor in Dance at The Ohio State University

BODY AND EARTH

BODY AND EARTH

An Experiential Guide

ANDREA OLSEN

Foreword by John Elder

Middlebury College Press   Published by University Press of New England   Hanover and London

MIDDLEBURY COLLEGE PRESS

Published by University Press of New England,

One Court St., Lebanon, NH 03766

© 2002 by Andrea Olsen

Printed in the United States of America

5 4 3 2 1

LIBRARY OF CONGRESS
CATALOGING-IN-PUBLICATION DATA
Olsen, Andrea.
  Body and earth : an experiential guide / Andrea Olsen ;
foreword by John Elder.
    p.   cm. — (Middlebury bicentennial series in
environmental studies)
Includes bibliographical references and index.
  ISBN 1-58465-010-9 (pbk. : alk. paper)
  1. Conduct of life—Problems, exercises, etc.
2. Human anatomy—Problems, exercises, etc.
3. Human ecology—Problems, exercises, etc.
I. Title. II. Series.
  BF637.C5 O49 2002
  304.2—dc21   2001008463

Front cover: Torii, white pine. Windsor, Vermont.

　　　　　　　Sculpture by Herb Ferris.

　　　　　　　Mud Portrait. Photograph by Erik Borg.

Title page:　Authentic Movement project. Photograph

　　　　　　　by Erik Borg.

Back cover:　Photograph of author by Bob Handelman.

This book is dedicated to my father,
    who taught me to love the land.
And to my mother,
    who encouraged a world view.

CONTENTS

MAPS

The first day of class, I tell my students we are beginning a journey. It is as dangerous as any unfamiliar terrain and equally disorienting, like exploring New York City would be for a rural person or camping in the desert for a city dweller. Very few of us know much about our body and its relationship to the earth. We need landmarks to guide us, maps on the journey as we learn its ways.

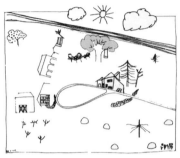

Map drawn by Anya Brickman Raredon, age nine.

FOREWORD

JOHN ELDER

In *A Sand County Almanac,* Aldo Leopold writes, "Land, then, is not merely soil; it is a fountain of energy flowing through a circuit of soils, plants, and animals." He invites his readers to recognize the life and unity of our planet, rather than to see only a conglomeration of raw materials. For our usual assumption that "soil" is just a "resource," one more commodity in the marketplace, he wants to substitute a more inclusive and participatory perspective on the "land." Such a shift broadens our sense of community and prepares for a more mindful and ethical relationship with the rest of the natural world.

In *Body and Earth: An Experiential Guide,* Andrea Olsen approaches the human body in a way that recalls Leopold's understanding of the land. The essential insight, for her as for him, is that human beings are included in the living circulation of the earth. "What is out there is in us," she writes, "and what is in us is out there." This is not just a metaphor for Olsen. She shows that, through becoming more attuned to the structure and processes of our own bodies, we also have the opportunity to register the balanced wholeness of the world more vividly. Such heightened awareness may move us past abstract concern for "the environment" to a more immediate and physical identification with the earth.

The Middlebury Bicentennial Series in Environmental Studies takes a bioregional approach. It assumes that nature and culture are best understood in relation to each other, and that their wholeness is most evident within the concrete specificity of a particular bioregion. Andrea Olsen contributes something new and valuable to such a project. She reminds us that a particular watershed must be physically entered, as well as enacted, through our work, our art, and the rhythm of our daily lives before it can become more for us than a mere idea. The body grounds this living "circuit" for her. The vital energy flowing into a human being from the surrounding landscape may surge back out into the world through mindful participation in natural processes.

In fulfilling its role as "an experiential guide," this book leads its readers through the conduits of the senses. The patterns that organize our own bodies are also found in the other organisms who are our evolutionary kin and in the living earth itself. The Gaia Hypothesis can be tested experimentally, like any other hypothesis, when a researcher systematically collects, refines, and interprets data through the evidence of his or her own senses.

The experiential core of *Body and Earth* is complemented by the centrality of stories to this distinctive curriculum. Standing at a fertile edge between dance and the environmental sciences, Andrea Olsen often expresses her insights through resonant stories of connection with nature. Tales of the vanished farmscape of her Illinois girlhood and of her family's annual migrations to Florida become performance texts through which her body can remember its former homes. They also become openings through which she and her readers, including the students who will use this book in environmental studies courses, may move more alertly into the landscapes they currently inhabit. Exercises and stories are complemented, in their turn, by the beautiful photographs associated with them. Such images help to assure that *Body and Earth* will be experienced on aesthetic and emotional levels as well as on an intellectual plane. In its pursuit of balance within diversity, this is a book that begins and ends in wholeness.

ACKNOWLEDGMENTS

I would like to honor my teachers and colleagues:

Janet Adler, for her insight into the discipline of Authentic Movement.

Bonnie Bainbridge Cohen, for her explorations in experiential anatomy.

John Elder, for his reflections on place-based ethics in college education.

Caryn McHose, for her articulation of evolutionary movement.

Anne Love Woodhull and Gordon Thorne for their commitment to art in the heart of community.

John M. Wilson for his cross-cultural perspectives in dance.

Portions of the text have previously appeared in *The New England Review* ("Farmstories"), *Orion* ("Notes on a Sense of Place in Dance"), *Contact Quarterly* ("Being Seen, Being Moved: Authentic Movement and Performance"), *Whole Terrain* ("Dance and the Environment"), and through *Station Hill Press* (*Bodystories: A Guide to Experiential Anatomy*). Appreciation to all the editors involved.

Iranian silver kneeling bull holding a vessel. Southwestern Iran, Proto-Elamite period, ca. 2900 B.C. The Metropolitan Museum of Art, Joseph Pulitzer Bequest, 1966. (66.173). All rights reserved, The Metropolitan Museum of Art.

Shayna Rae Peevy. Photograph by Bill Arnold.

ENTERING THE TEXT: ORIGINS

Sometimes it's necessary to go a long distance out of the way in order to come back a short distance correctly.

—Edward Albee, *The Zoo Story*

Stories of place in our time are often stories of migration. I am a dancer and writer who moved from the Midwest to the Northeast. Between uprooting and growing new roots, I criss-crossed the United States and circled the globe. Where I've lived and traveled affects the way I construct my view of the world and how I map it out for others, detail by detail. In our age of mass information, I am one of many who think it is useful for the reader to know something of those doing the writing, what they care about, and how they have come to their point of view. In my life, college teaching has been both source and resource for my work, and performance has been crucial to my way of understanding landscapes. And by landscape, I mean the natural and cultural rhythms of a place, including its soil and topography, plants and animals, air and water, people and their various creations.

I was raised on a farm in the heartlands of Illinois. Childhood amid that fecund terrain included an annual migration to the then-wild southeastern coast of Florida for the winter months. During these first eight years of my schooling, I knew two places as home and valued the journey between. My college years in the Midwest included summers on the Connecticut shore at the American Dance Festival with the inspiring New York–based "pioneers of modern dance." Graduate school at the University of Utah and the professional years that followed included performing throughout the western states, punctuated with overseas tours, providing a worldview. My first college teaching brought our young dance company east to Mount Holyoke College, and I've remained in New England ever since. I now teach at Middlebury College in Vermont and continue to perform and write. Movement and landscape, body and earth, inform all that I do.

My relationship with the sciences began with anatomy. Seventh grade with Mr. Barlow, then high school zoology with Mr. Green involved looking at the insides of things. This felt essential. When I joined the graduate program in dance at the University of Utah, my professor of anatomy and kinesiology, Dr. John M. Wilson, said, "All lasting dance techniques are based on anatomical truth," weaving, as he often would, science, history, and philosophy in a single phrase. I have now taught anatomy for twenty-eight years, deepening my knowledge of and fascination for this subject. Teaching anatomy as an artist makes me an amateur, not an expert, in science. It also

I wrote my first stories in a group led by Doug Anderson. A Vietnam medic, he had watched whole villages destroyed, held dying soldiers in his arms, seen children blown to bits. In his presence you didn't bother to write fluff. He encouraged me to be more specific, to make my images more concrete.

When I leave my writing group, I am often shaking. It takes me hours to settle down. It is not the spontaneous exercises that trigger my response; I feel nothing as I write. It is reading my work aloud. Amid this small group of peers, I never know what to expect. As the words unfold, I might feel energy choke my throat, tears come from nowhere, unaccountable emotion. I have come to enjoy these sensations, even in public readings, allowing the vibratory power of the voice to unleash its response.

allows me the role of liaison and translator between the sometimes impenetrable walls that separate disciplines in higher education.

One has to know a lot about a subject to speak simply; and often the more we know, the less there is to say. Instead, we embody attitudes and states of attention, communicating through our actions as well as our words. It has taken me eight years to refamiliarize myself with the distinct vocabularies of various sciences. I want, in this text, to retain the struggle of comprehension, something of the process of going from "outsider" to "insider." Based in specificity, the language of science can bring clarity or can confuse the newly initiated. Yet we each know the earth—the rocks, animals, plants, air, and water where we live. In approaching the sciences we can cultivate curiosity, humor, and awe, as we remember that every field of study begins and ends in mystery; the origins and the advanced edge are often conjecture, and "facts" change as new facts are discovered.

I had been a professional performer and educator for two decades when I moved to a wildlife sanctuary in western Massachusetts. My first book, *Bodystories: A Guide to Experiential Anatomy*, reflecting my years of work with the interior landscape of the body, had just been completed. At the sanctuary our focus was outside. My naturalist, botanist, and biologist mentors would set off, butterfly nets in hand, identifying dragonflies, searching for the unnoticed rare plant, and debating about invasive species. Just as with the body, there was always something to look at, something to look for, and a new place to explore. As well as preserving habitat for wildlife, the sanctuary's mission was to assist those of us who wanted to learn to see the world afresh. It was good-spirited fun, a safe environment for exploration.

Meanwhile, I continued teaching one semester each year at Middlebury and became caught up in the enthusiasm of the burgeoning Environmental Studies program. Interdisciplinary in approach, its success required participation by faculty across the curriculum. The students, with typical enthusiasm, drew their own links between disciplines and requested a course exploring the relationship between body and earth. Thus, I offered a January term experiment called "Ecology and the Body," which two years later became a course called "Body and Earth."

In this process, I was encouraged by gifted colleagues who spearheaded the Environmental Studies program: John Elder, professor of English, was an eloquent spokesperson for nature writers; Stephen Rockefeller, professor of religion and then dean of the college, illuminated the intrinsic intelligence of body and earth in his courses; and Stephen Trombulak, professor of biology, determinedly kept environmental issues in the eyes and minds of faculty, students, and community.

In many ways, my new course drew on the farm heritage of my childhood. I had been schooled to observe closely. As my father, a painter, focused on his watercolors, seated on a portable campstool in the field, his brush revealed the interconnections of branches and sky, roots and soil, shadows and sun. As my mother, a violinist, helped me pick the endless rows of green beans or arrange zinnias in a vase, I learned how to see both the detail and the rhythm of the land. My parents were educators from a lineage of educators. They helped me to understand the patience of discovering afresh what was right there in front of me all day long.

Body and Earth is written from and about the New England landscape. Encouraged by my editors to shape a text that "could be written in no other place," I was inspired to reduce my travels and stay home. It took concentrated effort, including research, writing, and daily sojourns on the land, to begin to know my New England home the way I know my body. Drawing on what I know best, the rigors and truthfulness of dance, I began writing performance texts about my triangle of homes in Vermont, Maine, and Massachusetts. Speaking as I moved required that I integrate the information embedded in the text into the expressive language of my body, an experience of inhabitation. What wasn't essential fell away. The stories in this book are perhaps the most personal, poetic aspect of the text, exposing the underlying current that moved me from beginning to end, to beginning once again. Completing each chapter, these short texts are intended to be read aloud, engaging the breath rhythms and emotional investment of the spoken word.

The book spans my seven years as a stepmother. Meeting boys of nine and thirteen and accompanying them through adolescence, with few community rituals supporting the process, is humbling. Parenting journeyed me back through my own development, finding the cracks and places of resistance and exposing them to light. Engaging with teenagers reminded me of the deep vitality, optimism, and potential of adolescence. It also highlighted the ways in which our entire culture is adolescent in its resistance to the process of maturation that could move us toward a new way of living with the land. As the boys ventured off on wilderness trips or their own creative explorations, I was invited to experience beyond what I had known. My view of life had been to *survive* adventure; now it was to grow a heart large enough to *encompass* what came.

This writing also spanned the death of my father. His passing changed my relationship to this American soil, linking generation to generation. Chief Luther Standing Bear, in his 1933 autobiography, says about his childhood on the Nebraska and South Dakota plains: "The man from Europe is still a foreigner and an alien. . . . Men must be born and reborn to belong. Their bodies must be formed of the dust of their forefathers' bones." Knowing my father's ashes were spread in the sea, I began to understand his words. Thus, the cycles moving from life to death, from daughter to mother, from visitor to inhabitant, form a defining arc to the progression of the text.

I have written *Body and Earth: An Experiential Guide* as an educator who dances, performs, and lives an involved life in New England with a husband, stepsons, and a dog. Each chapter or "day" includes elements I have found vital to my own learning: visual imagery, factual information, movement exercises, anecdotes, and personal stories offering diverse inroads

When selling our family home, my sister cleaned out my father's desk. Among his drawing pencils and photographs were four neatly folded pages, carefully preserved through the years: a high school essay. Handwritten script, etched neatly between blue lines, detailed how his father had deserted his family when he was twelve, how he vowed never to speak to him again, the fierceness of his feelings. Amid this highly personal writing, red circles surrounded spelling errors, corrected punctuation. On the outside, an F was imprinted in bold red ink. I wonder if my father ever wrote again.

We have so carefully learned to disguise what we think and feel. Psychologist Carol Gilligan writes that girls assume a voice at puberty and use it long after it has outgrown its effect—perhaps for a lifetime. Through academic training and the daily lessons of our lives, many of us have been taught to be good and to be quiet or to say what others want to hear. The feeling of risk that accompanies telling the truth is enormous. What is the purpose of writing, after all? To perpetuate what we already know or to open new doors?

into whole-body learning. The goal is to inspire you, the reader, to your own stories, maps, and ways of knowing. As we open our perceptual resources to the present moment, wherever we are, wherever our roots have been, we inspire care, investing our human resources in the land and communities we call home. Throughout the book, you are encouraged to take your own body seriously, to know it as familiar, as the vessel through which you experience the earth.

Pen, painting by Jim Butler. Oil on canvas, 27.5 in. × 90 in.

INTRODUCTION

We react, consciously or unconsciously, to the places where we live and work, in ways we scarcely notice or that are only now becoming known to us. . . . In short, the places where we spend our time affect the people we are and can become.

—Tony Hiss, *The Experience of Place*

Body and Earth is about relationship. It is a guide to help us reflect on what has shaped our attitudes and ideas about the world and about ourselves. Through our education in our homes, in schools, in our travels, and in our spiritual communities, we have developed a relationship to our bodies and to the environment. In this book, scientific views and experiential exercises are interwoven to help us investigate these attitudes and ideas, to notice if they are currently useful to our lives and to the health of the earth.[1] With attention to how memory and new information interact, we can become aware of our biases and assumptions and stimulate our imagination and curiosity for further exploration.

The body is the medium through which we experience ourselves and the environment. The ways we gather and interpret sensory information affect both how we monitor our internal workings and how we construct our views of the world. Thus, the text begins by exploring underlying patterns and perception, including basic principles of life and earth sciences that have shaped and currently affect our views of both body and earth. Next, we focus on the weave of relationships between bones and soil, breath and air, muscles and animals, digestion and plants, water and fluids. The text concludes with animating aspects of body and earth, including sensuality, creativity, and spirituality. Thus, the book provides a context to revisit and revision what we know, by noticing patterns of interconnectedness.

Throughout this study you are invited to choose a "place" in nature near enough to home for consistent field study visits. With notebooks in hand, we attend to specifics of our environment in order to connect to the places we live as participants and inhabitants, rather than as passersby. Weather patterns and seasons help us notice natural cycles of change, experience death or transitional stages as part of life, and deepen our relationship to the rhythms of the natural world. The goal, as we heighten sensual awareness of the environment and track our perceptual biases, is to expand our ability to respond, our response-ability, and ultimately to recognize ourselves as active participants in the world.

Language is a map for a way of thinking. Recognizing that language can point toward but does not replace experience, I have attempted to present specific information to enhance our creative explorations. In this study we

HUNTING FOR HOPE

Environmental writer Scott Russell Sanders tells of a backpacking trip he took with his sixteen-year-old son. His numerous books and articles detail the changes in his Midwest landscape, the destruction of ecosystems, the carelessness of our relationship with the earth. At the end of a difficult climb, the son turned to his father and said, "Your writings are destroying my hope." This comment inspired his next book, called Hunting for Hope.

learn the names of things as a way of claiming intimacy. We see not simply "leg" but "femur;" not "bird" but "barred owl." We use naming to suggest valuing through specificity. With a fresh view of language we can enhance our sense of familiarity, of family. Names reflect the universality as well as the uniqueness of each individual person, plant, or soil type. We are part of a larger group with shared characteristics, and we are also unique in each moment.

The art images, writing suggestions, and "to do" exercises in the book suggest forms for creative expression. They represent the life work of many individuals filtered through my own experience, offering a brief introduction to much larger fields of inquiry. In many cases the primary source of an exercise has been obscured as, over time, information has been passed from teacher to teacher. By using these exercises to enhance awareness of body and earth, we honor a heritage much larger than ourselves. Multicultural images encourage a world perspective, children's art refreshes, and illustrations clarify. Sometimes it takes many views to point toward truth.

Storytelling is included as an integrative process. The experience of reading, telling, and hearing stories can enhance our connection to people and to place. In many ways, storytelling transmits an experience so that listeners envision it for themselves. Re-creation in the form of storytelling is a kind of re-membering, assembling pieces to create a whole in which many dimensions can be experienced at once: emotions, ideas, facts, sensations. Within our bodies, a good story engages the limbic system of the brain, the area that registers interest and emotions, and engages memory.[2] It's the part of the mind that keeps one turning the page when reading a book. A good story touches all of us, brings us to the moment, not simply as observers but as participants.

Time is a consideration for the journey through *Body and Earth*. The overview of thirty-one "days" or learning sessions can be approached as a month of exploration, a twelve-week course meeting three times a week, or a progression to do at your own pace, alone or with a group. No matter what time is set aside for study, we are cultivating timelessness. Some experiences take a lifetime to understand; others, a decade; some are integrated in a moment. Building this quality of timeless exploration into our lives in a conscious way, helps develop the sensibilities essential to humane and creative living.

This book provokes an essential question: How best to live on this earth? By focusing on our human bodies as the vehicle through which we experience ourselves and the world around us, we learn to value the earth equally to the self. Thus, the perspective is both anthropocentric (human-centered) and ecocentric (earth-centered). The body and the earth are complex and profoundly interconnected entities developed through billions of years of evolutionary process. Our task is to develop a dialogue with this inherent intelligence—to learn to attend. By enhancing active awareness of our bodies and the places we live, we deepen our engagement in the intricate and delightful universe we inhabit.

When I ask students to write their history of place, many create eulogies for places that no longer exist. Even at twenty years of age, they see that the open field is now a parking lot, the favorite woods a subdivision, the family home long since departed. Writing about place is often writing about loss. We begin by recognizing our grief; we continue by celebrating connection to the places that shaped our lives.

Environmental studies students at my college are like the art students of the sixties. Energetic and hopeful, they feel they have a role to play in changing the world. And the results of the environmental focus are evident, changing the face and values of our college. Backpacks are everywhere; beards and hiking boots balance J. Crew; organic gardens and compost heaps replace overstuffed dumpsters; alert, lively senses counter the numbness of alcohol. We need reminders that change is possible and that each discipline, each person, has a role to play.

What *Body and Earth* offers, simply, is a map to guide experience. In editing the text, I've attempted to honor the body's capacity for absorption. In reading, one can only integrate what is relevant to the moment of attending. As writer, I offer what I can hold in one life, shared with friends like a good conversation. So when I had to choose what to include in a chapter, I would go outside and read aloud to the plants, go to the studio and speak as I moved, or go to the mountains and tell the story about water to the stream to see if it rang true. And that is what, finally, I offer to you in this experiential guide.

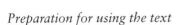

Preparation for using the text

Find an indoor space suitable for private work. A wood floor and natural light is ideal; the space should be warm and comfortable. Minimize distractions: unplug the phone, turn off the computer, remove your watch. Wear loose-fitting clothing; no shoes or socks. Have pencils and a journal or pad for your notes and drawings. Establish a realistic schedule; consistency of time and place helps calm the mind for focused exploration. You will also choose an outdoor place (See Day 1); each offers unique resources.

"To Do" movement explorations

Read through (and tape-record yourself if possible) the entire "to do" section first, then begin. As you become familiar with the process, it will become easier to follow the instructions. When working with a partner or in a group, have one person read aloud; then change roles. Respect your own limitations; modify any exercise that does not feel comfortable to your body. Use a yoga pad, meditation cushion, pillow, or chair for support as necessary.

Use the margins in the book to record your experience. Note what actually happens—your feelings, sensations, and associations—so that you can observe your process over time: "I was distracted" or "I felt light and relaxed."

Creative writing

Write from your whole body. Try to keep your pen moving, as you maintain a nonjudgmental attitude. Resist editing or censoring your words. The task is to let your body write you, that is, to become a vehicle through which your own stories emerge.

Read all of your writing aloud; include every word. When you read, both hear and feel your language. Often the experience of reading is different

Drawing by Claire Crowley, age two.

My stepson's seventh-grade math class focused on graphs. The teacher began by taking a survey of what students "were most afraid of." The final pie chart, colored in pale red, green, and blue, showed homelessness was the major fear for 60 percent of these rural students, followed by war and divorce. They were facing the fear that there is no place on this earth to call home. How do we acknowledge that each of us has a place; that both our body and the earth are home?

from writing. Cultivate a supportive listener within yourself. If emotions come, allow yourself to continue reading with the feelings. We need our whole self present in our words.

When giving feedback to others in a group, comment on what stands out for you, what *moves* you. Maintain your nonjudgmental but discerning mind.

I

Underlying Patterns and Perception

Cape—rope bridge. Photograph by Bill Arnold.

 # DAY 1

Basic Concepts

We are not separate from the earth; we are as much a part of the planet as each cell in our bodies is a part of us.
—Mike Samuels, M.D., and Hal Zina Bennet, *Well Body, Well Earth*

We begin in wholeness. The interconnectedness of human life with the world around us is the subject of our study.

Body, in this text, refers to all aspects of what it means to be human. The parts may sometimes be referred to individually as soma, soul, spirit, psyche, physical body, emotional body, intuitive body, energy body, thinking body, mental body, or collectively as person, self, "I." In our study, the word body is an inclusive term referring to the whole being.

Earth, in this text, refers to all aspects of our planet. The parts may sometimes be referred to individually as atmosphere, ecosphere, biosphere, hydrosphere, geosphere, mantle, crust, core, or collectively as Gaia, globe, world. In our study, the word *earth* is an inclusive term referring to the whole planet. Body is an aspect of earth.

Place, in this text, refers to a particular part of the earth that we know through direct experience in the body. Relationship to place is a process of assimilation, without which there can be no understanding. It is through our interaction with specific landscapes and environments that our movement patterns, perceptual habits, and attitudes have been formed. As we reflect on the places that have shaped our lives, we recognize that body affects place and place affects body in a constant process of exchange.

Where we focus our attention affects what we perceive. In this text we attend to the body as the medium through which we experience the earth. At various points in our exploration, we must put down our books, quiet our words, and simply go outside. Participation is the connecting link to awareness. As we open our senses to the natural world, we can recognize the experience at hand as the primary resource for our learning.

CONNECTIONS

When I first taught a course called "Ecology and the Body," a colleague met me at a department meeting with a stack of books and a typewritten note. The first sentence stated that ecology has nothing to do with the body. The note ended by saying that if there was no math in my course, it was not ecology. The Environmental Studies chairperson responded to my concerns with the reminder that scientists use language to be specific, writers to encourage association. Both are useful. He also quoted Rachel Carson: "You can't write truthfully about the sea and leave out the poetry."

Several weeks later a student reported a sign posted in a bathroom downtown: "Don't flush, save the Ecology." This helped me see the need for specificity in language; it also encouraged my sense of humor. Then, at a college-wide lecture, a distinguished environmentalist began his talk by announcing that every elementary student now knows that ecology means the interrelatedness of all living and non-living things. From my perspective, the human body is included.

yogic breath:
Inhalation lifts the spine and fills the upper body

Exhalation pushes the abdomen in toward the sacrum

Inner and outer awareness
10 minutes

Lying comfortably or seated, eyes closed:
• With each breath, feel or imagine the exchange between the outer environment of air around you and the inner environment of your body. The outer environment becomes part of the inner environment with each inhalation, and the inner environment releases to the outer environment with each exhalation.
• Gradually open your eyes. Notice whether you can remain aware of the sensations of breath while adding vision. This is the primary dialogue we will be engaged in throughout our work: the capacity to maintain inner awareness while attending to the outer world, a process of *inclusive attention.**

* *Some students who protest the use of chemical spray on blueberry barrens in Maine, scorn pesticides and fertilizers in the grain fields of the Midwest, denounce pouring raw sewage into streams, and bemoan the cutting of trees in the rainforest, do not hesitate to take Ritalin—"vitamin R" (to stimulate brain chemistry so that they get their homework done), Paxil (to slow down or feel calmer), Motrin (for torn or overused muscles), or Valium (to get to sleep at night after too much coffee and too many hours of computer buzz).*

What goes into the bloodstream enters the tissues, alters hormonal secretions, and affects the overall balance of the body. Why is interconnectedness important when talking about the migration patterns of the yellow rumped warbler but not the hormonal secretions of the thyroid gland; DDT threatening reproduction of the bald eagle because of thin shells on eggs but not the possible effect of Paxil on sexual function? We are odd creatures, we humans. We still don't recognize that what is out there is in us, and what is in us is out there.

Constructive rest
10 minutes

Lying on your back on the floor, in a warm, private place:
• Close your eyes.
• Bend your knees. Let your feet rest on the floor, slightly wider than your hips. Let your knees drop together to release your thigh muscles.
• Rest your arms comfortably on the floor, below shoulder height.
• Allow yourself to be supported by the floor.
• Allow your breath to move three-dimensionally in your torso.
• Allow the eyes to relax in their sockets, like pebbles dropping into a pool.
• Allow the brain to rest in the skull.
• Allow the shoulders to melt toward the earth.
• Allow the weight of the legs to drain into the hip sockets and feet.
• Allow the organs to release toward gravity.
• Allow your mouth to gently fall open; your tongue to relax.
• Feel the air move in and out through your lips and nose.

In constructive rest, rather than controlling your body, you let it be supported by the earth. As you release your body weight into gravity, the disks are less compressed and the spine begins to elongate. Constructive rest is an efficient position for body realignment. It releases tension and allows the skeleton and organs to rest, supported by the ground. Constructive rest is useful at any time of day but especially if done for five minutes before you sleep. The relaxation of the body parts returns the body to neutral alignment so that you don't sleep with the tensions of the day. Constructive rest is discussed by Mabel Todd in her book *The Thinking Body: A Study of the Balancing Forces of Dynamic Man.*

* *Inclusive attention* is a term used by Susan Gallagher Borg, director of the Resonant Kinesiology Training Program.

Place map

30 minutes

• Draw a map of a familiar place. Choose any place you have lived or visited that evokes strong feelings. Take time to fill in details and important landmarks. Consider pathways, boundaries, and orientation to light. Don't worry about the process of drawing; use symbols to represent areas of specific memory or meaning. 10 min.

• Write about the map and/or the place you've drawn. 15 min. Read aloud to yourself or a small group. Breathe deeply as you read, allowing exchange between the inner landscape of body and the outer landscape of place. 5 min.

Place visit: Finding your place

• Find a new place outdoors that you can visit each day. Look for an area that you can enjoy, within walking distance from your home or workplace and private enough that you can visit consistently, undisturbed, for twenty minutes at a time, engaging inclusive attention. Follow the place visit with ten minutes of writing in your field journal. Remember to allow time for direct engagement with place before writing and reflection, valuing experience as well as the language used to describe it.

Standing amid a swarm of mosquitoes, my flyfisher neighbor in Maine addresses the principle of interconnectedness even more simply: "You can't have the fish without the bugs."

Quarry, New Haven, Vermont.
Photograph by Erik Borg.

FARMSTORIES (1994)

We left the farm when I was twelve. I don't remember the details of departure. There must have been weeks of packing and a moving van. There was a huge barn sale, I am told, and good-byes to friends and neighbors. I don't remember anything. I don't remember what happened to the thirteen cats who lived in the outbuildings and supervised the field mice or who took or killed the last of the grumpy laying hens whose eggs I had so diligently gathered through the years. I don't remember the day that Rosebud, our five-gaited mare was driven away in the horse trailer or whether I kissed my best friend Cindy good-bye and promised to write and be friends forever. I don't remember emptying my bedroom closet of its pile of discarded dolls, seeming angry or sad about no longer being loved; whether Huffer Puffer, our very stupid but lovable cocker spaniel was returned to his original owners; or who took the trunks of grandmother's white dresses and linens that my mother had so carefully saved to pass on to her girls.

I don't remember anything about the move except arriving at our new house on Park Place and sleeping in the same familiar bed. I vaguely remember arriving; meeting Judy, my new best friend; taking Traveler, our collie, for his first walk on a leash to the park at the end of the street. But I don't remember saying good-bye to the fields, to the luscious cherry tree, which gave us fruit for jam, or to the garden where my mother grew zinnias and I came to know the fecund smell of overripe tomatoes. There was no farewell to the pump house where we stored our tadpoles; the deep cistern that we were not to fall into; the giant elm tree where the cicadas left their shells; the willow tree, which was our dollhouse; the outdoor stone fireplace where I prepared flower-petal soup for of all my imaginary children. Or the long flat view, which showed the horizon, and the houses of all of our friends and neighbors; and the line of dust warning that a car ten miles off was coming our way. I don't remember saying good-bye to any of this, and it is with me still.

Read aloud, or write and read your own story about home.

 # DAY 2

Attitudes

The failure to develop ecological literacy is a sin of omission and of commission. Not only are we failing to teach the basics about the earth and how it works, but we are in fact teaching a large amount of stuff that is simply wrong.

—David Orr, *Ecological Literacy*

Our attitudes inform our actions; the way we think affects what we do. One prominent view in Western culture is that nature is a "thing," an object with utilitarian value to be bought and to be sold. With this consumerist focus, we may consider the empty field, the lake, or the mountaintop to be property, a storehouse of resources, or a challenging landscape to conquer or control.

Another view offers a radical alternative: that nature has intrinsic value in and of itself. We can experience the world around us as an organic living thing. It is not object but subject. It has interiority, subjectivity. It has something to teach us, and it inspires respect. When we have this attitude, the natural world can evoke awe and astonishment, stimulating connection to the sacred, integrative forces of life.

The same attitudinal values are prevalent about the body. One predominant view, perpetuated by our educational and medical habits, is that the body is an object, a machine to be repaired when it breaks down. It is our property to do with as we please. It is a resource to get us from here to there, a commodity to help us get what we want, a storehouse of resources, a challenge to control, to conquer or overcome.

Hayloft, painting by Philip Buller. Watercolor on paper, 22 in. × 33 in. Collection of J. Slovinsky.

The radical alternative in body attitude would be that the body has intrinsic value in itself. It has interiority, subjectivity. It has much to teach us if we learn to listen. We can consider that we are part of a vast interconnected system, rather than separate from the world around us. We are nature too.

As we recognize our interconnectedness with the rest of the earth, we begin to see the world from a relational perspective, supported by our context rather than in isolation or at odds with it. We experience the air, the water, the soil, the animals, the people around us, and our own bodies as familiar. In this view, both body and earth are home.

Underlying the environmental crisis is a crisis of attitude. We are coming to the end of our nonrenewable resources, such as fossil fuels. We are also coming to the end of renewable ones, such as clean water, clean air, and healthy soil. This is true of the body as well. We can no longer count on resiliency in our structure: our bodies are too busy recovering, attempting to filter internal as well as external toxins. Yet we can consciously cleanse our systems and balance our energy, just as we can protect the resources around us. Our attitudes toward our bodies have everything to do with the health of the earth.

As we open our senses and attend to the earth, we encounter grief. Emotional memory is encoded since childhood, associated with specific terrain: the beach explored on family vacations, the tree climbed in moments of retreat, or the room where a child was born. All too frequently, loved ones are gone, places paved over—in reality or in our hearts. Earth systems, too, are dying. Forests, songbirds, and the air in our lungs are in trouble.

Grief is a natural response to loss. Engaging our grief requires softening our guard. The process includes several essential stages: acknowledging our denial, allowing emotion, accepting death as part of life, and expressing through creative forming—giving voice to the forceful feelings that occur. Our attitudes toward our grief affect our ability to feel and to act.

Environmental Studies is one of the fastest growing majors at universities across the country. Alternative health care is a growing force that challenges traditional medical practices to include a holistic view of the body. Books, magazines, and art reflect environmental perspectives and healthful incentives. There is no discipline, institution, or individual exempt from environmental concern and responsibility. As we become aware of our underlying attitudes toward our bodies and toward the places we call home, we experience the dynamics of living an interconnected life.

Any integrative experience is a spiritual experience, humanist John Dewey reminds us. One component of mystical or spiritual involvement that underlies ecological concerns is the feeling of being an integrated part of the whole planet. Aesthetic experience nurtures this sensibility in people. Thus, the arts have an essential role to play in encouraging us to face the unprecedented challenges of our time.

Our attitudes can be affected by information, inspiration, or direct experience. We can learn how a lake is polluted by reading the newspaper, seeing a painting, or finding a dying loon on the shore. Each instance connects person to place in a moment of alertness.

As we envision life, we create it. As we think, we do. Experiencing ourselves as participants, we foster attitudes in ourselves and in our communities that allow a lively and respectful dialogue between body and earth.

◻   ◻   ◻

Body map
30 minutes

• Draw a map of your body. Start with an outline and fill in the details. Include what you know and what you feel: anatomical information, ideas, emotional connections, colors, images, injuries, words, intuitions, memories. Follow whatever captures your imagination. 10 min.
• Now fill in the context around your body—the environment. 10 min.
• Write about your experience. Read aloud. Pause at the end of each sentence or phrase, and take a full breath—inhalation and exhalation. Allow a final breath cycle at the end of the story. 10 min.

With a group:
Using a large roll of wide paper, cut body-length pieces. In partners, one person lies on the sheet of paper with eyes closed; the partner takes a crayon or marking pen and traces around the outside of the body, in one long continuous line. Change roles. 10 min.
• Take the outline of your body and fill in the contents. Consider context, outside the outline, as well. 10 min.
• Write about your experience. Read aloud to your partner. 10 min.

Breathing spot (Child's Pose)
5 minutes

In constructive rest: Roll to your side, flex your arms and legs close to the body and continue to roll to a "deep fold" position: limbs folded, chest resting on thighs, knees spread slightly for comfort, forehead on the floor, spine curved.
• Rest the warm palms of your hands on your lower back, between pelvis and ribs. Through touch, encourage the movement of skin and muscles in this area as you breathe. On the in breath, your diaphragm is pulled down toward the pelvis, compressing the organs and expanding the back surface of the body as your lungs fill with air; on the out breath, the diaphragm releases (toward the ribs) and the muscles soften. We call this area of your lower back your "breathing spot." Encourage its movement with each breath. As breathing deepens, invite sensation to travel all the way down your spine, and spread into your buttocks and legs; then up to the heart, spreading to shoulders, neck, and skull.

In yoga, this is called Balasana, Child's Pose. Imagine a soft, rounded spine, responding to each breath. If you hold tension in your lower back, this is an excellent exercise to increase circulation and lengthen muscles.

Yoga: Child's Pose, Balasana.

FARMSTORIES: MODELS— THE TRAVELER

My mother was a traveler. She was always at ease with the wealthy. In 1939, after two years of college and before her first teaching position, she took "the grand tour of Europe" collecting the ideas and objects that were to fill my childhood. She sailed across the Atlantic, arriving in England, then in Africa, traveling on the Continent, encountering Mussolini, turning around, and heading for home.

On the farm, this heritage of world adventure was transmitted to us by the Della Robia Madonna and Child porcelain plate hanging over our kitchen doorway, by the leather gloves and amethyst ring from the Ponte Vecchio in Florence; the carved-ivory Coloseum on the bedstand; the transparent blouses embroidered by women in Dubrovnik, Yugoslavia; the woven vests and skirts from Norway.

Each of these objects, however small, was to detail a path that would make me a traveler. And so, in my twenties, when I began my own journeys, I walked under the smiling Madonna, wearing a transparent blouse embroidered by women who knew that life is how we embellish it, that's all we get. That life is the body. That's how we see it, smell it, taste it, and love in it.

Read aloud or write and read your own story about attitudes.

To write: Place story
2–4 hours

Write your history of place, using a chronological approach. Include all the places you have lived and visited. Consider home, travels, dreams, and longings. Reflect on the place-origins of your ancestors. Notice the ways your place history has affected your movement and your attitudes: did you grow up by water, near forests, or surrounded by city streets? How does place affect your life today?

Place visit: Body scan

Lying in constructive rest or seated comfortably, with a vertical spine and eyes closed, begin a body scan. Pass your attention part by part through the body, beginning with your face and skull. Notice any sensation that occurs on this part of your body. It might be an itch, a tingling, or the pressure of your body against the ground. Move your awareness to your neck. Notice any sensation on your neck at this moment in time: the touch of air to skin, heat or coolness. (Repetition of language helps to focus attention.) If you feel no sensation on an area, notice that, nonjudgmentally. Sensations are happening all the time, whether we are aware of them or not. Move your attention through every body part: the back surface of your body, front surface, sides, pelvis, each arm and each leg. Remember, if you feel nothing, just wait, inviting awareness; then move on. Finish by observing breath as it falls in and out of your nose and mouth, moving the ribs, muscles, and skin. Open your eyes, remaining aware of sensation. Is your attitude toward place affected by deepening attention to body? 20 min. Write about your experience. 10 min.

Body scanning helps to develop an equal relationship to all parts of the body, with no hierarchies or areas of avoidance. It is a component of Vipassana Meditation, a Buddhist practice also known as insight meditation.

Photograph by Erik Borg.

DAY 3

Underlying Patterns: Body

Evolutionary Movement uses the metaphor of evolution, from the state of undifferentiated wholeness, through cell, vessel, tube, fish, reptile, and mammal, to open up fluid movement capacity and inspire a sense of relationship with the natural world.
　　　　　　　　　　　　　　　　—Caryn McHose, *Resources in Movement*

We live in a fluid body; our origins were in the primordial seas. If we follow the evolutionary story, derived from fossil record and close observation, we see that our ancestors traversed through various body forms. Through all of these changes, we retained our liquid core. The amniotic fluid in the womb and our blood retain the saline content of the sea. Our skin keeps us from drying up; in fact, a human can be described as a sack of water walking around on feet! Although we may think of ourselves as solid, fixed, or hardwired, we are indeed fluid creatures, with adaptability and responsiveness as key characteristics for survival on earth.

Underlying human complexity is the unity of the single cell. The unique pattern for the whole body is contained in two strands of DNA housed in the nucleus. The fluid cytoplasm of each cell, like the body as a whole, is approximately 70 to 80 percent water. The selectively permeable cell membrane, like our outer skin, both separates and connects internal contents and

HERITAGE

My colleague, Caryn McHose, teaches the fluid body by introducing the image of "blenderized tissue." It's like putting the body in a blender and turning it on high, she suggests. Suddenly, all tissues are one tissue; we are back to the primitive cell.

Study for the Couple—Woman, painting by Gordon Thorne. Watercolor, 36 in. × 45 in.

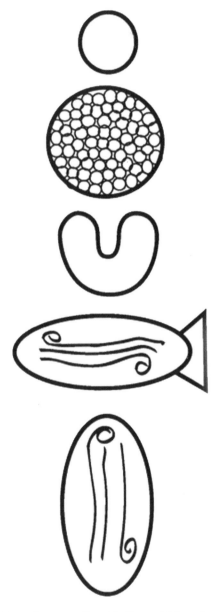

Top to bottom: cell; multicell; vessel; bilateral symmetry (horizontal): fish; bilateral symmetry (vertical): human

the external environment. A collection of like cells of similar structure and function is called a tissue. Groups of coordinated tissues form organs that comprise a body system. For example, muscle cells form muscle tissues, which make muscles, which create the muscular system. For our study we will differentiate seven body systems, based on the work of Bonnie Bainbridge Cohen and the School for Body-Mind Centering®: skeletal, muscular, nervous, endocrine, organ, fluid, and connective tissue.[1] Although we can look at each system individually, it is essential to remember that the body functions as an interrelated whole and that the systems balance and support each other.

We can revisit our phylogenetic history, the development of our species through evolution, to discern various body systems. From the oceanic matrix came the asymmetry of the first living cell in a shallow tidal pool, the sharing of resources in multicelled organisms such as sponges or sea coral, the vessel-shaped digestive cavity of sea squirts attached to the ocean floor, the central organization of radial symmetry in starfish, and the bilateral symmetry of fish, with many permutations in between. This oceanic heritage of our species is still present in our structure, such as the hollow tube of our digestive system and our segmented spine.

As our predecessors washed up on shores and became land creatures, gravity and inertia placed new demands on successful forms. The skin became the mediator between the fluid interior and the air-exposed exterior, modulating exchange of fluids and nutrients and maintaining a range of temperature suitable for life. Our ancestors traversed from belly slithering to four-footed to two-footed creatures. The head lifted away from the earth to heighten effectiveness of seeing, hearing, smelling, and tasting. What had been a mobile spine was now also used for stability, as the primitive tail and fins differentiated into legs providing effective locomotion. Eventually, in the bipedal stance, hands were free for carrying, manipulating, grasping, and tool use. Hand and visual acuity required coordination, learning, and memory and stimulated the nervous system, increasing brain size. These neurological connections still exist in bipedal hominids, supporting more complex connections such as contemplation, creativity, and imagination.

Human development, our ontological progression from conception through birth and early developmental processes, generally reflects phylogenetic history. Early developmental sequences of all vertebrates are similar, although there are deviations in timing. For example, arm buds from widely differing species are almost identical in the embryo, yet they may develop into a wing, a flipper, or an arm.[2] Within the human developmental progression, we have each passed through the stage of the first cell of a new individual, called a *zygote,* at the moment of union between sperm and ovum; a hollow multicellular ball called a *blastocyst* (implanted in the uterus around day 7); a three-layered *embryo* with central umbilical cord (around day 16), followed by the head/tail differentiation of bilateral symmetry (around day 24) and the budding and the emergence of tiny hands, mouth, and feet of the fetus (by day 56—eight weeks), initiating our venture into the complexity of human form.

All this occurs suspended in the amniotic sea within the uterus, supported by fluid and stimulated by the polyrhythmic sounds of the heartbeats, rumbling organs, and vibrations of external sounds (listen through a stethoscope!). Massaged by the mother's breath and rocked by the rhythm of her walking, the human fetus develops over 280 days (40 weeks) toward the process of birth. A significant shift in environment occurs as the newborn is greeted by the challenges of adapting to life in the world, in air, in gravity.

Bonding with the earth underlies all other developmental responses. A healthy baby bonds with air on the first breath, with earth by releasing weight to be held, and with mother by touch and nourishment by the first suckling. We have to release our weight down to the earth in order to lift the head up. We have to feel the ground to push away, to initiate movement, and to have support for a reach. Humans require connection to air, earth, and nourishment for survival, as well as touch, movement, and community.

Another of the many reflexes from our evolutionary heritage is *physiological flexion*, drawing the body parts toward center, and it is present in the womb. (Touch a caterpillar and watch it curl protectively toward center.) This is balanced by *physiological extension*—stretching outward—present during birth. The modulation between flexion and extension continues throughout our lives as we draw inward, returning to safety, and extend outward, daring to risk. *Head righting*, reflexively supporting and lifting the head during movement, protects the brain and allows focused perception of the environment. Just as bonding takes us toward the earth, head righting moves us out into space, supported by the ground.[3]

Reflexes and developmental patterns are encoded in the body for survival, supporting coordinated growth of all the body systems. These patterns essentially retrace evolutionary history through movement during the first year of life, preparing us for the complexity of our bipedal gait. To understand this process, we return to the work of Bonnie Bainbridge Cohen, who has articulated the patterns and their implications in our lives. *Condensing*

Athletes of all ages accumulate injuries from sports, reporting as many as thirteen surgeries by their college years. When we do a body scan, there may be no sensation. Why feel if it hurts? Numbness can be learned, as a defense. When we begin the "to do" exercises, there may be emotion: feelings about coaches who encouraged them to stay in games even when injured; irritation at their own bodies for not doing what they command.

Often, thoughts are conflicted because coaches were respected mentors; their bodies carried them to success. They began sports because of physicality and camaraderie; as competition took over, health was sacrificed for the win. Now they can't move. At twenty or twenty-one, they seek a new relationship with their bodies.

As sensation returns, the body can begin to heal. At first, there is discomfort: tension, bulk, strain. But eventually, the hypertoned areas relax, and a new conversation develops. Our structure responds to the task at hand; if you invite a sensitive alert body, you can have one. It's the heritage of our species.

Frog, drawing by Laura Lee. Graphite.

Painting, two Kangaroos. Oceanic, Australia, Northern Territory. Oenpelli bark, paint.
The Metropolitan Museum of Art, The Michael C. Rockefeller Memorial Collection, Bequest of Nelson A. Rockefeller, 1979. (1979.206.1514). All rights reserved, The Metropolitan Museum of Art.

and *expanding* are considered basic cellular movements in the human body: condensing establishes ground, connection to earth; expanding establishes spatial integrity, connection to air.

The interchange of push and reach patterns modulates the dialogue between self and other, essential for effective functioning with the environment and community. The *spinal yield and push patterns* draw the body in toward center, integrating head to tail and stimulating internal connectivity—orientation to self. The *spinal reach and pull patterns* extend outward to the environment, based on internal connectivity. Thus, the rhythm of yielding and reaching, internal connectivity and outward expression is the continuum we negotiate as human beings, basic to our existence, present in our movement patterns.

The necessary movement sequences of the developmental journey take us through rolling, sitting, crawling (on belly, homolateral), creeping (on hands and knees, contralateral), and kneeling to thoroughly integrate and prepare the neuromusculoskeletal system for the rigors of a vertical stance. This gradual (and overlapping) progression through movement patterns also supports essential fusion of the three bones of the hip sockets. When babies are allowed to pass through all stages of development before standing, rather than being lifted or placed on their feet, they are thoroughly prepared for the complexities of walking.

The transition to standing combines all of the patterns, initiated by a reach of the head and hands. Balancing the tone of front and back surfaces, along with pushing away from the earth to reach, stand, and walk through space, engages our dynamic dialogue with gravity—balance in the upright stance.

TO DO

Pouring the fluid body*
10 minutes

Lying on the floor, eyes closed:
• Imagine you are a single cell floating in the ocean, with fluid inside, suspended in fluid outside. Your skin is a semipermeable membrane, selecting what flows in, what flows out.
• Pour your fluid contents in any direction. Imagine you are totally suspended in a fluid, warm sea, moving and being moved with the tides.
• Now bring your attention to your skin, the selectively permeable membrane. Move with your focus on the skin, the outer membrane.
• Keep pouring and rolling, and bring your attention to the fluid contents inside the skin. Pour the fluid into particular body areas to initiate movement, like an ameba with a temporary protrusion of the protoplasm, a pseudopodium, that serves in locomotion or food gathering.
• Move your body slowly so that you can perceive sensations. In this asymmetrical movement, there is no up, no down, no head or tail, no right or wrong way to move; enjoy the sensations of wholeness and disorientation.

- Pause, noticing what has occurred. This state of nonjudgmental awareness is called "open attention," simply noticing sensations, emotions, thoughts, and images as they occur.
- Slowly add vision, remaining aware of sensations.

Vessel breath*

15 minutes

Find a comfortable position, seated, eyes closed:
- Focus your attention on your mouth. Start by yawning, stretching the mouth and back of the throat.
- Continue to allow the mouth to open and feel or imagine the sensation of stretch hollowing the center of your body, like shaping an empty vessel (a bottle or vase), with the pelvis as its base. (In humans, the trachea and esophagus separate for breathing and digestion, but imagine that there is simply one open chamber continuing down through the organs.)
- Imagine a primitive sea squirt attached to the ocean floor. Continue to breathe and notice if any sound emerges.
- Gently play, with a gradual opening of the lips, mouth, and throat. Allow a breath that is barely audible to emerge. As you continue opening the mouth and moving breath more deeply down the core of the body, it may feel odd, stimulating primitive patterns.
- Continue this breath for some time, following impulses for movement as they come.

This is related to Ujjai, a breath in yoga in which you slightly activate the vocal folds and surrounding tissues (glottis) deep in the throat to heighten sensation.[4]

Fish swish*

10 minutes

Lying in constructive rest, eyes closed:
- Focus attention on your spine (the 24 vertebral bones plus the sacrum and coccyx, connecting head to tail).
- Initiating with your tail, begin undulating the spine side to side on the floor, like a fish or a snake.
- Imagine a mouth on the top of your head, a tail extending from your pelvis. (Relax and "disappear" the legs!) Try initiating the undulations with your head, swimming toward or away, directed by the special senses in the skull.
- At some point, see how small the undulations can be; micro-movements may take on a life of their own, with subtle waves undulating your vertebral column as you breathe.
- Roll over; try the undulations on your belly. Allow your organs to hang off the spine, like a fish. Pause in open attention.

OF SPECIAL INTEREST: MORO REFLEX

The startle response, called the Moro Reflex (in developmental terminology), comes from our ancestors. When a warning is sounded in the jungle, our chimpanzee relatives take to the trees, using both hands and feet to climb and swing their way to safety. Babies on the mother's chest hang on for themselves. A strong reflex is encoded: to arch back, opening the hands and feet momentarily, so they can grab forward again, clutching big handfuls of fur to get a stable grip. We can see this pattern of release and holding on (extension followed by flexion) in our own startle response. When a car horn causes us to "jump," for example, a slow-motion view would show our spine arching, throwing the arms and head backward, rapidly followed by a protective closing of the front surface and a clutching of hands. Depicted in cartoons, the startle reflex is instantaneous, common to us all, encoded in our tissues.

Often, only the extension phase takes place, so the startle is not resolved by the hugging or taking-hold phase. Sometimes, when ongoing or intense trauma occurs, like war, abuse, or even the constant stress of work, we can find ourselves living in constant startle, or shock. The result is a rigid spine; unable to let go or hold on, we are frozen in time. Stimulating the front surface can be a useful intervention. This moves the focus of incoming information from the back (where the dorsal root enters the spine, bringing sensory information) to the front (where the ventral root of the nerve emerges, conveying motor activity—action). As we bring the information "to the front," we can begin to work with it, to understand and act. Sometimes this takes years; sometimes it can happen in a moment.

FARMSTORIES: WATER

I know myself to be one with water. How did I find this to be true? Perhaps it was being pulled below the surface of the sea and thinking I would never return; submerging a tired body in a bath, recovering; tossing a backpack over a desert waterfall, jumping; seeing the wet world of birth, after so many dry words describing.

I know myself to be one with water. When did I find this to be true? It wasn't during grade school or junior high, or high school or college, but years after solidifying my image of self, when someone told me, in a voice that I could understand, that we are mostly water, with a few minerals and fibers to hold us together. That all living things are mostly water. That the earth is mostly water.

I know myself to be one with water. The daughter who assists her father in his dying, by withholding fluids; the mother who assists her son in his leaving, by withholding tears; the son who assists his love to her life, by withholding semen; the grandfather who frees his children from his curse, by withholding spittle: each act made from love helps us drink deep from the well of life.

I know myself to be one with water. Wading, standing waist deep, immersed.

Read aloud, or write and read your own story about water.

Fish swish (with a partner)*
10 minutes

One partner lying in constructive rest, eyes closed; the other standing at their feet, knees slightly bent:
• As standing partner, reach down and encircle your arms around your partner's legs, below the knees. Slowly and carefully, begin walking backward, giving a small side-to-side "swish" to your partner's pelvis, noticing how the spine undulates in response. Encourage relaxation of your partner's neck so that the undulation can move throughout the spine. Work at a speed that is safe, so that there is no whiplash or strain!
• Pause, release the knees, and take a moment in open attention; change roles.

Place visit: Attention to underlying patterns of body

Seated, eyes open or closed: Bring your attention to breath. Explore the vessel breath, encouraging the receptivity of the body. Initiate small undulations in the spine as you sit at your place, remembering the fish swish. Imagine the fluid contents of your body responding to the environment around you: the air, water, plants, animals, and soil. Remember, they are filled with water too! 20 min. Write about your experience. 10 min.

*The exercises in this section are drawn from Caryn McHose, movement teacher and educational bodyworker. Also see Caryn McHose and Kevin Frank's "The Evolutionary Sequence: A Model for an Integrative Approach to Movement Study," *Rolf Lines*, May 1998.[1]

DAY 4

Underlying Patterns: The Upright Stance

Our senses, after all, were developed to function at foot speeds, and the transition from foot travel to motor travel, in terms of evolutionary time, has been abrupt.
—Wendell Berry, *An Entrance to the Woods*

Bipedal alignment is our two-footed stance, a high center of gravity over a small base of support. As our vertical axis constantly sways over our feet, reflexive contraction of the muscles of the lower legs keeps us on balance. A subtle shift past the base initiates walking, striding, or running. In effect, we are constantly falling; instability is basic to our structure. Walking, arms swinging freely by our sides, is an underlying rhythm of our species.

In the evolutionary story, our quadripedal mammalian forebears emerged during the Age of Reptiles, around 180 million years ago. For these small, ground-dwelling insectivores and herbivores, larger carnivorous animals made life threatening. Some species moved to the trees and adapted to a posture of hunkering on branches. In this squatting position, feet grasped the branch and legs were folded (flexed) to the belly, leaving the hands or forepaws free for eating, grooming, and gesturing. The hunkering posture encouraged structural changes in the body: the heel bone (calcaneus) migrated to the back surface opposite the toes, an arch was formed by the connecting bones of the foot, and the pelvis shortened to allow the femurs to fold in squatting position.

The new use of the hands also developed a cross-pattern of thumb opposite fingers for grasping. Picking the plentiful nuts and flowers of the evolving seed-bearing plants (angiosperms) required coordination of eyes with hands. Our ape ancestors began hanging from their arms and swinging from limb to limb through the treetops, called brachiation. This allowed extension of hip and shoulder joints, repositioning of the shoulder bones (scapulae) to the back surface of the body for hanging and lateral reaching, contralateral rotation at the waist necessary for swinging, and elongation of the spinal curves. The curve in the lower back (our lumbar spine) formed last, after the arboreal swinger returned to the ground as a semi-quadruped. The transition then to upright posture was accompanied by the anterior (forward) curve of the lumbar segment. These characteristics prepared the way for two-footed, vertical posture: a bipedal stance.

Multidimensional agility of body and brain evolved simultaneously in our structure. Without moving our feet, humans can reach in front, to the sides, or behind. This three-dimensional rotation of the spine and hips permitted quick response in any direction. The free-swinging pelvis held the

Nancy Stark Smith and Andrew Harwood.
Photograph by Bill Arnold.

STANDING UP

"The Small Dance" is a movement exercise developed by dancer Steve Paxton. You close your eyes and stand with your weight balanced in vertical alignment. The intent is to notice all that is happening inside in what we call stillness: the myriad shifts, micromovements, rhythms, and idiosyncrasies that happen moment by moment. The small dance is our dialogue with gravity and the dynamics of the earth. As we open our eyes, we can remain aware of this conversation and "listen in" throughout the day as we stand in a line, talk on the phone, or walk through the woods.

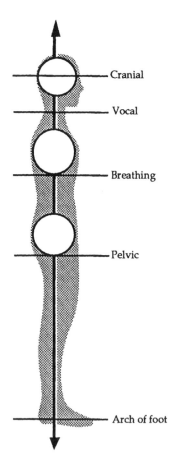

Three body weights with horizontal diaphragms.

organs in a lightweight bowl and allowed mobility (unlike our close relative, the gorilla, who evolved from a common ancestor but whose large pelvis restricts vertical extension and whose knuckle walking and vegetarian diet encourage a more passive existence). The use of tools and the development of family groups increased the need for articulation and communication. In essence, we could stand, move three-dimensionally, manipulate our environment, and articulate with gesture and sound. Neurological adaptability increased our ancestors' potential for survival.[1]

The basic characteristic of our species, *Homo sapiens* (wise man), is the increased capacity of the brain. As all of our systems developed simultaneously with our skeletal-muscular changes, our physiological capabilities were matched by our capacity for three-dimensional thought in the past, present, and future. (Imagine our brain in the body of a horse.) We are able to reflect on where we have been, contemplate where we are, and plan where we are going. This capacity for reflection, planning, and manipulation of our environment brings the responsibility of choice. Our ability to plan and to shape our environment makes us responsible for what we create and for how we choose to live in that creation. As we reflect on the precarious stance of the human—center of gravity vertically balanced over the length of our feet, constantly falling, essentially alert—we recognize that responsiveness and responsibility underlie our relationship to the earth.

The evolutionary progression to verticality offered additional implications for our bodies. As three primary tissue layers differentiate in the developing embryo, the ectoderm becomes the skin and nervous system, responsible for communication; the mesoderm becomes the skeleton, muscle, and connective tissue, responsible for support; and the endoderm becomes the organs, essential for breathing, digestion, and reproduction. In a simplified drawing of the bilateral fish body, the brain and nervous system are on the top for perception and communication, the segmented skeleton is next for support, and the organs hang by connective tissue (mesentery) off the spine for metabolism and reproduction, creating an efficient layering in horizontal alignment.

Shifting these layers into human verticality alters our dialogue with gravity. What used to be horizontal support (i.e., our backs) is now a vertical series of balanced curves. What used to hang (our organs) is now stacked. What used to communicate (our spinal cord and brain) is now an axis integrating head and tail, sky and earth. Our genitals, belly, breasts, and face, once protected by their horizontal relationship with the earth, are exposed, creating a vulnerable but dynamic and expressive body. Communication within community is now face to face, belly to belly, responding to information from all the senses.

In the core of the body, several musculotendinous partitions, called diaphragms offer horizontal support in vertical alignment. Beginning with the skull, the cranial diaphragm creates a sling for the brain, cushioning it from impact as we walk. In the neck, a vocal diaphragm supports the structures for voice. In the torso, the thoracic diaphragm creates a mobile rhythmic floor for the lungs and heart and massages the organs during the process of breathing. The pelvic bowl includes a fibrous pelvic floor, offering resilient support for digestive and reproductive viscera. Even the arch in the foot, a webbing of muscles, tendons, and ligaments below the bony bridge of tarsal and metatarsal bones, creates a horizontal diaphragm, offering spring and shock absorption for our striding gait.

Along with the vertical axis supporting bipedal directionality, human structure includes three globes (skull, ribs, and pelvis) and four rotary joints of shoulders and hips through which we occupy spherical space. Although we often think of ourselves as flat and two-dimensional, as reflected in photographs, mirrors, and on television or computer screens, we have sculpted fullness. Curves and angles give force and agility to the body, connecting us with forms beyond and inside us. Nerves inform us about the landscape, interior and exterior. The human form inhabits spherical space, as multidimensional and diverse as our curiosity allows, inviting investigation and discovery.

In an efficient vertical stance, the skull, thorax, and pelvis are balanced around an imaginary plumb line or vertical axis. If we draw three ovals, representing these three body weights, and connect them with a vertical plumb from the top of the skull to the feet, we have a diagram of postural alignment in the body. The front is also called the anterior or ventral surface, and the back is the posterior or dorsal surface, as we look at the body from a lateral, or side, view. The three body weights connect with a series of reversing curves: anterior at the neck, posterior at the ribs, anterior at the lower back, posterior at the pelvis (sacrum), and anterior at the small curve of the coccyx—the ancestral tail—connecting to the front surface of the body through the webbing of the pelvic floor. Each spinal curve touches but does not pass the plumb line in front. The center of gravity lies in the area behind the belly button at the front of the (lumbar) spine, intersecting the plumb line. To locate the center of gravity (c.g.) in your own body, imagine the point created by the intersection of the three primary planes of the body dividing you equally by weight: front to back, side to side, top to bottom. The c.g. is in front of the fourth or fifth lumbar vertebra in most bodies.

Additional bony landmarks for efficient postural alignment include the two "feet" of the skull (occipital condyles) as they pass the entire weight of the structure (some 13–20 pounds) to the first vertebra of the spine (the atlas). The ischial tuberosities (sit bones) are the two "feet" of the pelvis, which point straight down in efficient postural alignment, standing or seated. Balanced alignment of the pelvic bowl requires that the front rim of the pelvis (pubic bone) support the organs. Imagine that the entire front surface of the body—pubic bone, belly button, sternum, and nose—has an upward energy flow, and the entire back surface of the body—back of skull, back of ribs, and back of sacrum—has a downward energy flow (like a waterfall), creating a cyclical pattern throughout the structure. The two body surfaces—anterior and posterior—complement and run parallel to the imaginary plumb line, or central axis.

When our bones are balanced in efficient alignment, we feel less. In fact, with minimal feedback in the nervous system, we may feel as though we

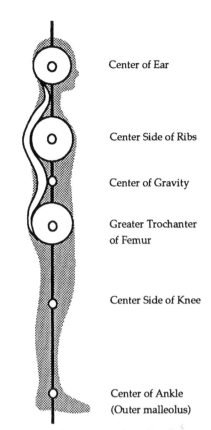

Center of Ear

Center Side of Ribs

Center of Gravity

Greater Trochanter of Femur

Center Side of Knee

Center of Ankle (Outer malleolus)

Three body weights with spinal curves and landmarks for postural alignment.

When my father was dying, I visited ancestral homelands: Samso island off Denmark. Standing in a cemetery filled with headstones marked Olsen, I felt myself amid family. Less than two centuries earlier, someone departed this land to plant roots in America; someone stayed behind to till familiar soil. Now, we share a common history of place. Bicycling amid rolling hills surrounded by sea, I realized that the adjectives depicting this landscape were those I had heard, for years, describing my dancing: lyric, gentle, remote. I felt as though this place-heritage inhabits my body, informing the way I stand, speak, stroke my hand through my hair, even though I had never been here before.

TO DO

aren't doing enough. People move to feel themselves working. But no sensation is the sensation of ease. As weight passes toward gravity through the bones, there is levity in the soft tissues, an upward release. Balancing the bones removes compression on other systems. Our spinal cord and nerves remain free for receiving and expressing; the digestive organs, lungs, and heart can function effectively, maintaining homeostasis in the body.

Alignment is relationship, to self and the environment. Even when standing "still," the earth is always moving, we are always moving. A gift of our bipedal structure is our multidimensional agility of body and mind, which keeps us alert and responsive, adaptable to change. The risk is that instability can cause fear and rigidity in our attitudes and stance. When we have a relaxed, toned body, if we are pushed, we recover. If we are pushed when rigid, we fall off center. Our fluid body, undulating spine, and reflexes of face, gesture, and language support our vertical and vulnerable selves.

Postural alignment (Mountain Pose)
5 minutes

Standing in plumb line, eyes closed:
• Touch the bump on the outside of one ankle. This is your *outer malleolus,* the lower end of the leg bone and the first landmark in postural alignment.
• Touch the large knob on the side of your upper leg bone (the part that touches the floor when you are lying on your side). This is called the *greater trochanter,* the second landmark in postural alignment. Line the greater trochanter directly over the center of your ankle.
• Touch the *center side of your ribcage,* the third landmark in postural alignment. Line it directly over the greater trochanter and outer malleolus.
• Touch the *center of your ear,* the top landmark in postural alignment. Lift your elbows to the side and imagine lines from each pointer finger meeting in the center of your skull. Do a small "yes" nod around this horizontal axis. Line this joint up with the other landmarks for postural alignment.
• Touch the top of your skull and imagine the plumb line extending upward and downward, creating a vertical energy line around which the body parts are organized. This is called "extended proprioception" (imagined sensation).
• Check that the knees are not locked; weight should pass through the center side of each knee joint.
• From postural alignment, slide into your favorite "hang out" posture; feel this position. Then, beginning with the feet (your connection to the earth) slowly reorient your body toward postural alignment. Repeat.
• Sometimes it is helpful to use a mirror to "check" alignment. Stand with your side to the mirror; align the body, feet to head; then rotate the skull to look in the mirror and notice the vertical relationship of body parts.
• In efficient alignment, weight passes through the bones, your mineral body, to the earth.

In yoga, the vertical stance is called Tadasana, Mountain Pose. Imagine a favorite mountain as you stand, allowing its qualities to inform your body.

Spinal undulations (standing)

5 minutes

Standing in postural alignment:
• Begin a spinal undulation from side to side, a "fish swish." Imagine a mouth and eyes on the top of your head, a fish tail at the end of your spine, like a trout or shark.
• Swing your tail (pelvis) side to side to propel the spine and head, or swim the head through the water to lead the spine and tail. Keep the movement side to side, as though your front and back surfaces are between two flat panes of glass. This fish pattern, evolving 400 million years ago, still lives in your spine.
• Change to a spinal undulation front to back. Still imagine the mouth and eyes on the top of the head, but now you have a flat tail, like a whale or dolphin. This mammalian pattern, appearing 180 million years ago, still underlies your movement.
• Change to a spiral pattern of the spine, unique to humans. Begin rotating the pelvis and allow all the vertebrae to respond, like a flag wrapping around a flagpole. Let the eyes finish the spiral, looking behind you as you twist. This pattern was present in your earliest hominid ancestors in Africa 5 million years ago, and underlies your multidimensional agility—the capacity to move in any direction with ease.
• Reverse the spiral until you have a full swing, wrapping left, wrapping right. Include the whole spine.
• Repeat each undulation, noticing any place there is holding in the body: the neck, the heart area, the lower back or tail. Encourage mobility of the spine in all three directions to support the stability of your vertical stance.

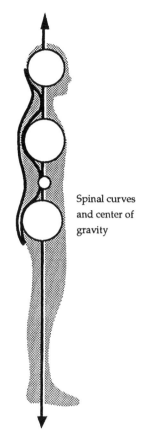

Spinal curves and center of gravity

Three body weights with spinal depth.

Photograph by Erik Borg.

Place visit: Attention to bipedal alignment

Begin standing in plumb line, eyes open: Look at the trees and plants around you from this vertical stance; imagine you are growing roots from your feet into the soil, intersecting with those of the plants at your place. Remember that the root system of a tree is often as large as its crown. Try spinal undulations standing; notice the mobility under your bipedal stance. Pause and observe place in open attention. 20 min. Write about your experience. 10 min.

FARMSTORIES: ARRIVING

When my parents were young, they decided to pull a long green trailer across the United States in search of a place to call their home, raise their children, lead a good life. They had passed through thirty-nine states and were headed toward California, when they found themselves parked next to the Grand Canyon with both children crying, "We want to go home." So they turned the long green trailer around and headed back to Illinois, to the farm my mother's father had given her as a present. He never expected her to live there.

When they arrived, they parked the trailer in the back pasture, under the black walnut tree, and set to work. Farming was familiar. Things were done just about the same as when my father's father had tilled the soil in Denmark. When we left nine years later, everything had changed. We had entered the era of more: more land, more money, more equipment. And of less: less community, less intimacy, less humor. But a photo of this first season shows my parents standing side by side in a field of corn, children by their sides, watching the sun set on the horizon. This was their place, their work, their time to learn and grow together with the land.

Standing, read aloud or write and read your own story about the upright stance.

DAY 5

Underlying Patterns: Earth

We have forgotten what we can count on.
—Terry Tempest Williams, Breadloaf Writers' Conference, 1997

It takes both a macroscopic and microscopic perspective to understand place. Macroscopically, we can reflect on basic occurrences over the past 4.5 billion years that have affected the Earth's land forms, temperature, and capacity for sustaining life, as well as ongoing influences such as the eruptions of volcanoes, recomposition of the atmosphere, and the movement of tectonic plates. We can consider patterns of rhythmic variation—rhythmicity—that affect relationship to place moment by moment, such as the Earth's rotation, which contributes to the cycles of day and night, weather, and the tides. Fluctuations in electromagnetic fields, gravitational pull, light and sound waves, and air pressure trigger such diverse responses as the migration of the monarch butterfly and the opening and closing of the night-blooming

RHYTHM

The concept of microscopic and macroscopic perspective can seem distant. But a film called Powers of Ten, made by the husband and wife team, Charles and Ray Eames, in 1977, offers direct experience. It begins with two figures lying on a picnic blanket on the shore of Lake Michigan in Chicago; then in ten frames to the second, we move by steps out into the universe until we are out of the galaxy. Then we reverse the process, returning in dizzying speed to the couple on the lawn. But then we go further: through the skin and inside the body into the cells, electrons, and empty space— as far in as we were out, to the power of 10. It stretches our imagination to watch this ten-minute film. I have seen it many times, and each viewing leaves me breathless, as it exposes the macrocosmic and microcosmic patterns that underlie our lives.

Sangam Ritual Series, sculpture by Michael Singer. Aspen Art Museum, Aspen, Colorado. Photograph by David Stansbury.

When we went through the Grand Canyon by raft, we passed from recent time to distant past and back in seven days—a journey of three billion years at least. The river was the connecting link, moving us through time. When we stopped at Redwall Cavern, a natural overhang (some 500 feet wide and 150 feet deep), our guide pointed out a crinoid "sea lily" stem fossilized in limestone, resting on the fine sand beach (400 mya). Tracing its curves, I reflected on spiral patterns present today in the chambered nautilus, with its mathematically exact proportions, and also in the Milky Way galaxy, in a single strand of muscle fiber, and in our DNA. I felt the delicate beauty of an early attempt at efficient form, a rhythmic pattern amplified through time.

datura, as well as influencing our human moods, biological rhythms, actions, and interactions.

Understanding place also requires a microscopic view to discern the primary building blocks that comprise our universe. Scientists have postulated that the universe is made of molecules (groups of atoms), atoms (made of particles), and subatomic particles held together by such basic forces as gravity and magnetism. Since the general acceptance of quantum mechanics in the 1940s, the distinction between matter and motion has been blurred. Particles, or waves, sometimes described as energy fields, constitute all aspects of our universe as far as we know, disrupting any notion of fixed and solid matter. Stars, trees, dogs, the paper of this page, and our bodies are composed of the same fundamental particles. In this view we can recognize the interconnectedness of cosmological community: humans are inextricably linked with the universe as an unfolding evolutionary process. As Carl Sagan reminds us, "We are made of star stuff. Our whole being depends on the universe."

Big patterns are both frightening and hard to perceive. Part of the pleasure or terror involved in engagement with the natural world can be a feeling of insignificance as we recognize ourselves to be a small part of a much larger whole. Stories have been created throughout human existence to help explain the origins of life and our place in the cosmos. Models reflecting religious, philosophical, poetic, and scientific views are available, although the scientific is dominant in contemporary Western education. Based on posing questions, observation, and testing, scientific ideas are presented as theories

Creation Myth, drawing by Micha Sam Brickman Raredon, age seven.

that change when new facts are discovered. Sometimes, several theories exist simultaneously until one is proven more accurate. It is useful to remember that at the advanced level of every discipline, there are more questions than answers. Our individual insights may provide new perspectives on the stories that shape our lives.

According to contemporary astrophysicists, an explosion about 4.6 billion years ago formed an interstellar cloud of gases (mostly hydrogen and helium) and dust (containing carbon, silicon, oxygen, iron, and other elements) that eventually formed our solar system. Within this swirling cloud, gravitational forces attracted dust particles from the dense gases and combined them to form solid rocks, building up massive globules called cold planetisimals. Thus, about 4.5 billion years ago, according to the geological record, the planets of our solar system were forming.[1]

The extremely hot proto-sun at the center of this swirling mass began to radiate heat and light into the galaxy. The planets close to the Sun, including Earth, were literally baked by this "external heat engine." The Earth's earliest atmosphere, created from the release of gasses from its interior due to heat and chemical reaction, lasted only a short time because of a weak gravitational field. Gradually, Earth retained a suitable atmosphere to modify temperature and allow the first life forms to develop. The chemical composition of this 40–60 mile thick gaseous envelope has evolved several times, affecting and being affected by life forms on the surface. Light from the massive fire of the Sun travels 93 million miles before arriving at Earth's surface—a journey that takes about eight minutes. Gravitationally bound to the Sun, planet Earth rotates on its axis every 24 hours, circles the Sun every 12 months, and orbits the center of our galaxy, one orbit every 220 million years.

It is speculated that the Earth was initially composed of the same materials at all depths. However, gravitational collapse and the disintegration of radioactive materials produced an "inner heat engine," which warmed the Earth and possibly created oceans of molten rock. As the liquid, boiling-hot surface of molten rock gradually cooled and formed a crust, the heaviest materials, like molten iron, were gravitationally drawn to the center. Through time, the Earth developed differentiated layers, unique chemically and mineralogically from inner to outer. This included a solid iron inner core and a liquid iron outer core (that together constituted one third of the mass of the Earth), a partially molten mantle, and a light rock crust, with no continents or oceans. The oldest rocks that have been found on Earth, offering evidence of these changes, are dated at 4 billion years.

Energy flows naturally from hotter areas to cooler areas, and volcanoes erupted and spewed their molten contents, including gases, solids, and liquids, full of chemicals, up to the surface. One theory suggests that this process changed the chemical composition of the atmosphere from a mixture of hydrogen and helium gas, held in by the Earth's gravitational field, to a mixture of gasses that included water vapor, carbon dioxide, and nitrogen. It is also postulated that meteorites, partially composed of chemically bound oxygen and hydrogen, bombarded the Earth, releasing these chemicals as water vapor and increasing the water resources of our planet. High up in the

TIME LINE:

4.6 bya	Explosion and interstellar cloud of gases
4.5 bya	Earth forming
3.8 bya	First water
3.4 bya	First life, photosynthetic bacteria (single cell)
1 bya	First sexual reproduction
750 mya	Multicelled organisms (cell colony)
500 mya	Hydra and starfish (radial symmetry)
400 mya	Bony fish, first land plants and animals (bilateral symmetry)
350 mya	Amphibians and insects radiate; coniferous trees
280 mya	Reptiles radiate
250–200 mya	Pangaea and Panthalassa
225–65 mya	Dinosaurs
180 mya	First small mammals and birds appear (reptiles rule)
135 mya	Angiosperms (flowering plants) widespread
65 mya	Continents in current positions
50 mya	Early primates radiate (hunkering)
22 mya	Early apes (brachiation)
5 mya	Hominoid/hominid (ape/human) split
3.7 mya	Homo erectus (bipedal alignment)
2 mya–10,000 ya	Glaciers (widespread 65,000 ya)
1.8 mya	First stone tools
130,000–75,000 ya	Homo sapiens

Dates indicate longer, overlapping periods of time; they also change with new findings. bya, billion years ago; mya, million years ago; ya, years ago

The first time I went snorkeling forever changed my view of life on land. There, below the undulating surface film of the Caribbean, was an underwater world as complex and diverse as any I had known. Trigger fish with their tiny mouths, colorful parrot and butterfly fish, angelfish with vertical stripes, serious-looking grouper and schools of ephemeral bonefish moved through the waters. Amid the living corals, sea plants, and algae, undulating sting rays, giant sea turtles, creeping starfish, and sand dollars inhabited this unique terrain. Later I learned that familiar mountains and valleys are also present in the sea floor, a vast expanse covering two-thirds of the globe. In fact, John McPhee reminds us, in Annals of a Former World, *that what is topsoil becomes ocean floor, and what is ocean floor becomes topsoil in a continuous rhythm of change.*

Earth's evolving atmosphere, it was cool enough for water vapor to condense into liquid water; it is possible that rain occurred over millions of years, creating the first oceans. Sedimentary rocks, marking the earliest presence of water on the earth, have been dated at 3.8 billion years.

Molecules of many chemicals were washed out of Earth's surface rocks, and shallow pools eventually filled with diverse groupings. One theory suggests that, when conditions such as temperature were right, a unique combination occurred that included molecules called amino acids; another theory postulates that amino acids came to Earth as a component of meteorites. Whatever their origins, amino acids became part of complex molecules that could make copies of themselves, and these molecules were incorporated into cells that could then replicate.

The elements that compose amino acids and cell structure are actually part of larger cycles. Carbon, nitrogen, and phosphorus are crucial to the existence of all living things. Carbon is stored primarily in deep oceanic sediments (and is essential for building physical structure and storing energy); nitrogen is present in gaseous forms in the atmosphere (and is a large component of many enzymes that break down carbon compounds and release energy for the body's use); and phosphorus is bound in the continental crust, where it weathers and becomes available for uptake (for the building blocks of DNA and cell membranes). To exist, all organisms depend on consistent ratios of carbon, nitrogen, and phosphorus cycling through the land, water and atmosphere.

Cells are the structural building blocks of all living beings. As the first living organisms, they exhibited the basic characteristics of life: the ability to reproduce, metabolize, and respond to changes in the environment. Earliest life forms were single-celled prokaryotes, similar to modern-day bacteria. The geological record shows that, about 3.4 billion years ago, self-feeding (autotrophic) prokaryotes, like blue-green algae, evolved; they could make their own food through photosynthesis, a metabolic process using carbon dioxide, water, and sunlight to form simple sugars for energy storage. From these unique, single-celled, photosynthetic bacteria, more complicated forms evolved, such as plants that could also make their own food, converting sunlight to organic matter by photosynthesis and releasing oxygen. This process once again changed the chemical balance of the atmosphere.

According to the fossil record, bacteria were the only organisms for the first two-thirds of Earth's history, dating back over 3.4 billion years ago. They reproduced exact copies of themselves by cell division. Sexual reproduction, combining genetic material from two parents, began around 1 billion years ago. Multicellular plants and animals first appeared 750 million years ago. The first land plants and insects evolved around 400 million years ago. The first birds and mammals developed over 180 million years ago. It has been 65 million years since the dinosaurs disappeared. Glaciologists tell us that over the past 2 million years and as recently as 10,000 years ago, large glaciers covered the North American and Eurasian continents and smaller glaciers occupied alpine valleys. Evolutionists remind us that hominids have

Marine protozoa, radiolaria.

walked the Earth for a mere 5 million years, with anatomically modern humans, *Homo sapiens,* appearing in Africa around 130,00 to 75,000 years ago. Reflecting on underlying patterns of temporal evolution offers a perspective of other than human scale to our present experience of place.

The theory of plate tectonics, or continental drift, has been generally accepted since the 1960s, suggesting that Earth's crust is divided into about twelve continental plates that float on the partially molten magma of the mantle and move with convection currents. It is proposed that 400–500 million years ago there were continental blocks (paleocontinents) whose movements and collisions are recorded in mountain belts. Although the specific locations of these early blocks are conjecture, fossil records and rock compositions help geologists configure their possible locations. Geologists propose that some 250 million years ago tectonic plates collided to form a single massive continent, called Pangaea, that stretched from pole to pole, surrounded by a universal ocean, Panthalassa. Around 200 million years ago the supercontinent separated into land masses that would eventually become the northern continents (Laurasia) and a southern supercontinent (Gondwana).

The northern Atlantic ocean was formed around 180 million years ago, when the northern continents formed a crack, followed by sea floor spreading, which separated Eurasia and the Americas. The southern Atlantic ocean was created around 130 million years ago when South America broke away from Africa. By 65 million years ago, India was connected with Asia, and Australia separated from Antarctica, producing planet Earth as we know it today. Plates continue to collide, affecting mountain ranges above and below sea level, with results such as volcanoes and earthquakes. They also slip past one another, as in the San Andreas fault in California, resulting in bedrock drop.

Glaciers once covered most of North America, and many of the features of our current landscapes are products of glacial activity. Over the past 2 million years and as recently as 10,000 years ago, glaciers advanced and retreated several times. They wore down mountain ranges; deposited large rocks (glacial erratics), pebbles, sand, and clay; and created lakes, ponds, and deltas. Around 65 thousand years ago glaciers covered nearly 17 million square miles of Earth's surface, and sea levels were more than 400 feet lower than today. Land bridges connected previously separated areas, supporting migrations to new territories.[2]

Musician Mike Vargas speaks of rhythm as "when things happen." He explains that "it has to do with the timing of events within a given time frame: there is rhythm in a moment and there is rhythm across a span of hundreds of years." Consider the placement in time of a cough or of the periodic eruption of volcanoes along the Pacific Ring of Fire. "Listen to the sounds around you for ten minutes," he suggests. "Note when they happened, for how long, and how they interact in time: the on-goingness of a waterfall, the chirp of a cricket, a blast of a car horn." In his view, rhythm is everywhere in a landscape, including the words on this page.

Bonding with gravity

10 minutes

Lying on your back in a comfortable position, eyes closed:
• Let your weight be supported by the Earth. Notice any part that seems to be "hovering" weightlessly above the surface. Try to soften or melt all of your body toward gravity.
• Now lift your head about an inch off the ground, feel its full weight and relax it back to the Earth.
• Lift a leg off the ground; feel its weight; relax it to the Earth.
• Lift your pelvis off the ground; feel its weight; relax it to the Earth.
• Lift an arm off the ground; feel its weight; relax it to the Earth.
• Notice your full body weight resting on the ground.
• Begin slowly to pour the contents of your body toward one side and roll onto this surface (bring your arms along). Feel your weight drain into the Earth.
• Continue to roll slowly, pouring the fluid contents until you are resting on the front of your body. Release your "bellital surface" into the ground.
• Roll, very slowly, onto the remaining side, pouring your contents.
• Return to your back and nestle your whole body into the ground.
• Slowly begin the transition to standing, remaining aware of your fluid contents moving with the pull of gravity.
• In plumb line, remain in open attention, noticing any sensations (thoughts, emotions, images) that occur. Add vision and continue awareness of gravity.
    Bonding with gravity underlies all other movement patterns. We must be able to release our weight down in order to push away, stand, walk.

Timeline

45 minutes

On a large piece of paper, alone or with a group:
• Draw a timeline of the history of Earth, from the origin of this planet to the present day. It does not need to be an actual line; consider other ways of representing time. Include information you remember from your studies and details from textbook sources.
• Find a scale that spans millions and billions of years, as well as recent history.
• Add three dates that are unique to your area of study or interest: the birthdate of Martin Luther King, the date humans landed on the moon, or the first performance of *The Rite of Spring*.
• Write about the origin of your version of the evolutionary story: teachers, films, books, church lessons, photographs, talking to friends, family discussions, visits to museums, television shows, computer games.
• Read your story aloud to yourself or a small group; show (and revise) your timeline.

De-evolutionary sequence

45 minutes

Begin standing in vertical alignment, eyes open:

• Let your plumb line fall forward to initiate a walk, like *Homo sapiens*, moving with multidimensional agility 130,000 to 75,000 years ago (ya).

• Explore your earliest hominid heritage in the African savannas, sharing food in family groups, before the erect stance, 5 mya.

• Reach your arm up and grasp a branch with your hand, brachiating like our early ape ancestors to locomote through the treetops, 22 mya.

• Pause and squat, hunkering on a branch along with other early primates. Feel the arch of your foot grasping the tree branch, eyes and hands free to pick food and groom and to communicate with others in your community, 50 mya.

• Travel down to the ground, like a shrew or other small mammal, 180 mya.

• Move through the land like an early dinosaur, using your big tail, 225 mya.

• Crawl or creep on your belly, like your reptile ancestors, 280 mya.

• Slither like your amphibian ancestors, moving between water and land. Imagine a salamander, needing a wet environment for survival, 350 mya.

• Swim off into a muddy pool, out to the ocean, leaving behind the first land plants and insects. Feel the undulations of your bony fish spine, 400 mya.

• Explore the mobile radial symmetry of the starfish. Attach to the ocean floor like the hollow vessel of the sea squirt, feeding as water flows in and out, 500 mya.

• Imagine yourself part of a cell colony, unique but interconnected, like a coral or sponge, 750 mya.

• Return to the integrity of a single cell, like a photosynthetic bacterium floating in the ocean, responding to sunlight, 3 billion years ago (bya).

• Imagine yourself participating in the fluid matrix before the cell, the raining down of the first ocean waters, 3.8 bya.

• Reflect on the erupting of volcanos from the molten core, changing the chemical composition of earth and atmosphere, 4 bya.

• Consider the differentiating of layers as Earth formed, the cooling of the crust, 4.5 bya.

• Consider yourself part of a swirling gaseous cloud, one of the tiny particles of an exploded star forming the basic building blocks of our universe, 4.6 bya.

• Pause in open attention. Write about your experience. Read aloud, feeling the ground as you speak.

Place Visit: Attention to underlying patterns of Earth
30 minutes

Walking to your place, eyes open, let your arms swing freely. Notice the rhythm of your walk and the contours of the land. When you arrive, stand in vertical alignment, eyes closed, and notice postural sway. Imagine the surface of the Earth moving under your body, the planet rotating on its axis, the solar system revolving around the Sun. Open your eyes, but keep your imagination active. Attend inclusively to the sensations of moving body and moving Earth. 20 min. Write about your experience. 10 min.

FARMSTORIES: SOIL

Each July when I was a child, we would drive the thirty miles to the Illinois State Fair and catch up on the farming news. We would drink fresh-squeezed lemonade, ride the double ferris wheel, visit our favorite displays, like the giant cow carved from butter, and walk through the animal barns. But in the hot and heavy afternoons we would go to the farming tents and try to judge the quality of a product by the personality of the salesman. We were given hats with their names, rulers and pencils for the kids, and pamphlets about fertilizers, hybrid seeds, pesticides, and herbicides. We chose Van Horn Hybrids, wore their hats, remained family friends.

Into our dark, rich soil we placed the fertilizers, eliminating the need for the fields of alfalfa to rest the earth. Into our dark, rich soil we placed the herbicides, eliminating the need for the carloads of high school students earning summer money by pulling the weeds from the corn, tending the crops that would fill their plates in winter. Into our dark, rich soil we placed the hybrid seeds, eliminating the rows of waving seven-foot-tall corn, replacing them with shorter stalks, more ears, denser rows—increasing the yield, decreasing the prices. The story is familiar. Rachel Carson told it well. Lick the tip of your finger, and on your tongue you will find many chemicals developed since World War II, most of which we know very little about.

It is was when writing about the dark, rich soil of Illinois that I found tears in my eyes. "There is no more topsoil in Kansas," my neighbor tells me. Hearing that there are twenty thousand insects for every square foot of soil cheers me up. There they are engorging, aerating, resuscitating the earth. I'm not really a lover of insects, but the idea that some species takes soil seriously makes me happy.

Place your bare feet on the earth. Feel the ground beneath you. Know the connection.

Read aloud, or write and speak your own story about soil.

DAY 6

Underlying Patterns: A Bioregional Approach

Place is security, space is freedom: we are attached to the one and long for the other. . . . What begins as undifferentiated space becomes place as we get to know it better and endow it with value.

—Yi-Fu Tuan, *Space and Place: The Perspective of Experience*

A bioregional approach merges nature and culture; humans are considered part of, rather than separate from, the natural world. A bioregion is generally defined as an area with biological integrity, including all interacting life forms. Environmental educator John Elder elaborates, describing a bioregion as a naturally defined landscape, comprehensive in both topographical and biological ways and also in the way it includes human culture. The resulting permeable boundaries may have little relationship to political borders and can be viewed on various scales from local to global. Dynamic edge zones, called ecotones, result where one bioregion overlaps another, creating particularly rich habitats that support life from both regions.

To know your bioregion, it is useful to identify specific characteristics used to determine bioregional boundaries. We begin with a geological overview, reflecting the movement of tectonic plates and the volcanic and glacial

NATURE AND CULTURE

My dance studio at the college has one wall of Vermont granite—metamorphosed limestone from warm, ancient seas. The floor is maple, cut in Vermont forests; the light-hued ceiling is pine. The piano, with its ebony veneer, reflects distant lands. I am aware at every moment of all that supports the body, gesture by gesture. Zen centers in Japan were modeled from ancient barns; this room reflects both. The dark-stained floor reminds us of simplicity, encourages us to touch our foreheads to the earth again and again. Conversing in this space, I ask a dance student what part of the country she might choose for graduate studies. She responds that place is not important. But I disagree. The landscape shapes who we are and whom we will become. These rooms, these floors, the mountainous horizon through the window are now part of everything she does. Each swoop of her arm holds a history of this room where we stand.

Canoe by Stephen Keith. Photograph copyright © Benjamin Mendlowitz.

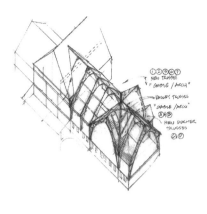

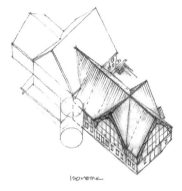

Barn reconstruction for Bramble Hill Farm, Amherst, Massachusetts. Frame isometric by Tris Metcalfe, architect.

activity that has shaped the terrain and waterways, influencing soil types. We also consider the resulting changes in climatic and atmospheric conditions, such as temperature and weather patterns. We then look at plants and animals, flora and fauna, and the ways they have altered with the arrival of humans. We reflect on the first Paleo-Indians, later migrations, and population centers. As we study agriculture, forestation, and industrialization practices, we can also consider prevailing religious, scientific, and cultural attitudes that affect stewardship of place. In other words, we look at soil and landforms, air, water, plants, and animals, including humans.

For practice in developing a bioregional perspective we will focus on Middlebury, Vermont, the home landscape of this text. Middlebury is part of the Champlain bioregion, one of six major regions in Vermont. Flanked on the west and east by the Adirondack Mountains and Green Mountains, respectively, the town is nestled in the fertile Champlain lowlands adjacent to nearby Lake Champlain, giving this area a four-weeks-longer growing season than that of the mountains of southern Vermont. The Champlain bioregion is part of the larger Greater Laurentian bioregion, which includes most of New England and extends north into Canada.

Geological history shaped the patterns that contour our contemporary landscape.[1] The Grenville orogeny is considered the first major mountain-building era of the Appalachian chain, occurring 1.3 to 1.1 billion years ago (bya). Geological records suggest that the oceanic crust bordering the eastern coast of a much smaller North American continent slid into the earth's interior, in a process called subduction. As the continental plates, moving with convection currents from the planet's internal heat, slowly collided, they created a lofty mountain range (comparable in height to the Himalayas)—the ancestral Adirondacks, which covered New York and Vermont.

You can touch rocks from the Grenville era today in the spine of the southern Green Mountains (Killington and Pico peaks) and the basement layer of the Adirondacks, although the mountains themselves gradually eroded and disappeared over hundreds of millions of years. The still-jagged Adirondack Mountains visible today are thought to have been formed from a geological "hot spot," erupting under the crust around 2 million years ago and followed by erosion. Thus, they are considered "young" mountains in geological time.

Around 590–550 million years ago, Vermont was near the equator, partially submerged under tropical waters, with a hot and steamy climate. No flora or fauna had yet moved onto land. Three collisions between the North American plate and the African plate occurred between 590 and 250 million years ago, shaping and reshaping the land and endowing this bioregion with its present contours.[2] The first collision, known as the *Taconic orogeny,* occurred around 450 (470–450) million years ago. Convection currents in the earth's mantle reversed and began to close the proto-Atlantic Ocean, causing an underthrusting of the coastal slab. Shoving coastal rock inland, the collision upfolded another majestic north-south-trending mountain range, now called the Berkshire Hills and Taconic Mountains. This period created most of the dominant mountains in Vermont.

The second collision, the *Acadian orogeny*, occurred around 400 (450–345) million years ago, as land masses again collided, refolding the Green Mountains. A part of the crust from the proto-Africa plate, which had rafted out through the oceanic crust and had formed a micro-continent called Avalon, eventually collided with proto-North America.[3] After the collision, as convection currents reversed, the continents separated and this part of Africa was left behind, newly attached to Vermont. If this geologic story is correct, New Hampshire and eastern New England were once African soil. Around 320 (345–280) million years ago, the final collision between the North American and African plates resulted in the *Alleghenian orogeny*, primarily affecting the formation of the Appalachian Mountains in the southern United States. Together, these mountain-building events—the Grenville, Taconian, Acadian, and Alleghenian orogenies—formed the Appalachian chain, extending from Quebec to Georgia. The elegant peaks were ground down by erosion and glaciation into the landscape we know, including the pastoral contours of the 350-million-year-old Green Mountains of Vermont.

By the end of the Alleghenian orogeny, around 250 million years ago, the supercontinent Pangaea had formed: a giant picture puzzle of continents with Vermont at its center, still near the equator. Broadleaf forests now covered the land and reptiles were abundant, including a plethora of giant dinosaurs. In the New England region, fossil remains show that three-toed dinosaurs strolled along what is now the Connecticut River; two-foot-long crocodiles that walked on four long legs and galloped with all four feet off the ground at once, left their 212-million-year-old fossilized remains in both Connecticut and Scotland.[4] Pangaea itself was encircled by a giant ocean, Panthalassa, filled with burgeoning life forms, including sharks and sea turtles—successful species that continue today.

When the supercontinent began to break up around 200 million years ago, the continental plates slowly drifted on convection currents to their locations around the globe, entailing significant changes in climate and in flora and fauna. The North American plate came to rest in the Northern Hemisphere, and Vermont developed a near-Arctic climate and a barren tundra-like landscape. Dinosaurs were widespread on every continent and in many varieties. Mammals began to appear around 180 million years ago, identified by fossil remains with small, shrew-sized jaws found in the western United States, Europe, and South Africa. Angiosperms dominated plant life, with flowers and seeds providing plentiful food. Mammals radiated, helping to pollinate and transport seeds in symbiotic proliferation.

The Ice Age began in North America around 2 million years ago, perhaps as a result of a collision with a meteorite, and ended as recently as 10,000 years ago. Four different glaciers descended from the north, repeatedly scouring the land, sculpting valleys, removing topsoil, and shoving the earth ahead of them as they traveled. The last, the Wisconsin Glaciation, encompassed all of New England in a one-mile-thick ice sheet, covering even the high peaks of the Adirondacks.

As the climate warmed, the ice sheet began to melt and retreat northward. Glacial residue includes piles of rocks and sand (terminal moraines);

"My family is in business," a student said. "I came to school to be an economics major; my goal is to make money. Someday if the environment is really bad, I'll be able to live where there's clean air, clean water." And he continued, "That's why I came to school in Vermont, where the environment is still pure." When we studied air, he recognized that the prevailing winds in our region of Vermont often blow in from the Midwest, bringing pollution from the factories in Cleveland and beyond; tree rings show nuclear fallout from tests in New Mexico; Vermont has serious problems with acid rain; and the state has one of the highest breast cancer rates in the country. He began to realize that there is no place on earth where money can protect you against pollution. The world is interconnected in such subtle ways that we must each attend to the whole.

When I teach yoga, we stand in Tadasana, Mountain Pose. The room grows silent. We feel bodies erect, weight dropping down, mineral bone meeting mineral earth. Eyes focus past the windows at the Green Mountains nearby, 350 million years old. Inner gaze connects to the skeletal core of our bodies. The breath is full, so all surfaces of the skin move as one. Time slows down.

north-south scratches in granite bedrock from the moving ice (striae); and giant boulders (glacial erratics) that now sit amid farmers' fields and in local forests. Glacial potholes and lakes reveal indentations in rock and soil from the massive weight.

The melting ice sheet traveled as far north as Burlington, Vermont, damming the northern drainage and spreading a giant Lake Vermont over its compressed roadbed. The lake was more extensive than today, and telltale markings of water levels can be found along the sides of cliffs and near the edges of the Adirondacks. The tops of nearby mountains (including Snake Mountain and Mount Philo), were tiny islands amid a vast waterway.

As the ice continued to move northward, the St. Lawrence River linked Lake Vermont to the Atlantic Ocean. Salt water flowed in, forming a shallow inland sea. Whale and seal bones, fossilized and found in Charlotte, Vermont, are dated from 10,000 years ago. As silt at the delta and the gradual rebound of the earth closed the northern outlet, the connection to the sea was lost, and the waterway once again became a lake. Freshwater Lake Champlain was much larger than its current size, with a rich, fertile valley to the east. The Champlain lowlands today feature a nutrient-rich soil of clay and silt from this time.

Many waterways in Vermont follow the northward path of the retreating glaciers. Lake Champlain itself, 125 miles in length, is sometimes referred to as the sixth Great Lake; it flows north to the Richelieu River in Canada, then to the St. Lawrence River and to the sea. Most rivers that flow into Lake Champlain, such as the Otter Creek, Lamoille, and Missisquoi, flow north, as does the Mad River further east.[5] The Connecticut River, forming the eastern border of Vermont, is an exception. This 407-mile-long river has its headwaters in small lakes of northern New Hampshire and travels south to meet the ocean at its mouth in Connecticut. In the 1800s and early 1900s this river, like others, was a dumping ground for industry and sewage, and people avoided the unsightly view; dams were built to generate power, diminishing the animal and fish populations. In fact, the Connecticut River was once described as "the most beautiful sewer in America." Now it is being restored, at considerable expense, as a healthful source of beauty for those who live near or visit.

Around 30,000–12,000 years ago human ancestors crossed the Bering Strait onto the North American continent (although stone tools found in the Southwest suggest much earlier human inhabitation).[6] Small bands of Paleo-Indians arrived in the Champlain bioregion around 11,000 years ago, hunting in a still glacial environment. In a mere fifty years, the giant woolly mammoths became extinct, most likely due to overhunting and climatic change. Eventually, giant buffalo and woodland caribou were hunted in hardwood forests; plus walruses, seals, and whales in the new Champlain Sea. Through the years native peoples went from small nomadic bands to hunters and gatherers with specific territories and seasonal migrations, followed by a gradual shift to farming with the domestication of plants and animals. Artifacts can be found throughout the bioregion, detailing different periods. During the Woodland era (3000–400 ya) there were five language families; the Abenaki tribe in the Champlain bioregion spoke Algonquian, contributing place-names such as Lake Memphremagog and the Missisquoi Rivers, which generally describe some feature of the waterways for other travelers.[7]

By the time the French explorer Samuel de Champlain and his men arrived by water in 1609, native peoples were well established in the area, and had coexisted with the land successfully for thousands of years. In epidemics of 1616 and 1633, 80–90 percent of the Abenaki were eradicated due to white man's diseases (to which natives had no immunity) and from fighting (including battles with their neighboring Iroquois enemy). From the 1600s to 1776, ongoing battles, including the French and Indian War, kept European settlers away. With the French defeated in 1763, the English arrived in earnest, lured by "cheap" land and "untouched" resources. It took as little as fifty years for white settlers to overrun indigenous peoples and cut the virgin forests. European farming techniques reconfigured the landscape into fenced fields and pastures. Immigrants flocked to the region, and the remaining Abenaki "disappeared" into the culture. As recently as the 1900s, texts reported that there were "no native peoples" in Vermont, the Abenaki having merged so thoroughly into the communities; yet there are over two thousand self-proclaimed Abenaki today in Vermont and one thousand in neighboring Quebec.

The arrival of Europeans decimated animal life as well. The beaver, key to healthy drainage and water purification across the nation, was a prime target for attack; their pelts "bankrolled" early colonists, as they could be sold to Europeans for use in felt hats. Author Alice Outwater, in her fascinating book *Water,* describes how English entrepreneur William Pynchon was granted a monopoly on fur trade in this region. By the mid-1670s, nearly a quarter of a million beaver pelts had been shipped to London from the Connecticut River Valley alone! Felt hats made from beaver fur were worn by both sexes throughout Europe and eventually in the Americas. By 1700 even the industrious beaver was extinct along the eastern coast, at a great loss to the water drainage of the region.

By 1812, Vermont was the fastest-growing state in the union; woodlands were burned for potash and charcoal for iron mills and logged by farmers clearing land for sheep farming. By the 1840s, Vermont was 20 percent forest and 80 percent cleared, and the state was an ecological disaster—erosion had depleted the soil, and all large mammals were locally extinct, including whitetail deer, beaver, catamounts, bear, caribou, elk, mountain lions, timber wolves, and wolverine. After the Civil War, in 1865, Vermont became the slowest-growing state in the country. Whole communities of men who had gone to the war never returned. Homesteads near the Canadian border were farmed by French Canadians who came in to tend the farms. The underground railroad once had way stations in Vermont and Maine en route to Canada, but only a few African Americans settled in this region after emancipation. Indeed, the end of the Civil War marked the beginning of the migration westward and the resulting displacement of native peoples.[8]

Sculptor Isamu Noguchi reminds us, "We are nature too."

As communities of hill farmers headed out for the newly opened Midwest, they abandoned depleted farms, leaving cellar holes and stone fences as markers. This exodus marked the slow renewal of the Vermont bioregion. Hardwood forests grew again in abandoned fields, and in the early 1900s a second cutting occurred, followed again by reforestation. Today, Vermont is 80 percent forest and 20 percent cleared, with a mixed northern hardwood forest containing sugar maples, oak, spruce, fir, beech, red and white pine, butternut, hemlock, and white birch. The bioregion offers an example of recovery, along with the memory of land—plants, animals, soil, water, and people—desecrated by careless stewardship.

Wildlife has returned. In fall and spring, migrations of hundreds of thousands of snow geese stop to feed and rest on their north-south journey. Although the passenger pigeon, once present in such large flocks, is now totally extinct, other birds have returned: wild turkeys, peregrine falcons, eagles, osprey. As you walk through the woods, you might encounter a moose, bobcat, whitetail deer, coyote, or fisher; it is even possible to imagine a catamount or wolf venturing down from Quebec. Stone fences, arrowheads, and glacial erratics evoke earlier stories of the land.

Humans are part of the landscape, contributing to biological exchange. Within the narrow span of time that Europeans and immigrants from all over the earth have settled on this North American continent, each bioregion and the continent as a whole have been altered. Humans now inhabit the terrain on a large scale, as visitors or residents. It is time to understand ourselves as co-inhabitants with the land and learn to tell the bioregional stories of the places we call home.

☐   ☐   ☐

TO DO

Place scan

20 minutes

Seated or lying in a comfortable position, eyes open:
• Bring your attention to *soil* and *rock*. Where is there soil in this room or place? Use what you can see and what you know, engaging extended proprioception. If you are indoors, take time to consider the possibilities within this place and nearby, considering metals, glass, sand, and rock. Be sure to include your own body and other humans.
• Bring your attention to *air*. Where is there air in this room or place? What can you smell, taste, feel, hear, or see concerning air? Remember, the air is a medium of travel for birds, insects, the spores of plants, chemicals, sound waves, light waves, and humans. Consider what is happening with the seasons and with the sun and moon. Bring your attention to any aspect of air that you can perceive at this moment in time, in this unique place and season.
• Bring your attention to *water*. Where is there water in this room or place? Consider what's happening with the water cycle, the water table, in the watershed for your region. Remember your body is mostly water.

• Bring your attention to *animals,* our relatives on this earth. Where are animals in the place? Notice the season, the effects on animals; the food you have eaten being digested in your stomach, insects and tiny microorganisms.

• Shift your attention to *plants.* Where are plants visible in this room or place? Even if there are no plants in sight, the floors may be wood, plants are being digested in your stomach; we wear plant fibers. Notice or imagine trees, grass, and plants of the season, contributing to the oxygen sustaining your life.

• Pause in open attention. Try scanning, eyes closed, with senses heightened.

Harley. Photograph by Leight Johnson.

Telling your place story
10–20 minutes

Seated or standing in a comfortable position, eyes open (alone or with a partner):
• Describe your place out loud to a partner or an imaginary person. Feel your connection to the ground and your fullness of breath as you speak. Be direct; try to not squirm or avert your eyes. Add as many details as you can about the specifics of your place, talking for five minutes.
• If working with a partner, change roles. Discuss your experience.
• Then tell the story of your bioregion, talking for five minutes. Notice light, air, objects, and people, orienting yourself to the place where you are speaking. Change roles and discuss.

Place visit: Orientation
30 minutes

Find a new route to travel to your place. Walk somewhere you've never walked before, expanding your awareness of the context of your place. Include a place scan as you walk, attending to earth, air, water, plants, and animals, engaging inclusive attention. At your place, locate and face east (the direction in which the sun rises). Pause and take two full breaths. Face south, west (the direction the sun sets), north—breathe in each direction, taking time to locate and feel each orientation. Then extend one arm up to the sky, one down to the earth, orienting your body as a connecting link between sky and earth. Finish your experience by facing toward your home, acknowledging the direction and the connecting pathways. Repeat the place scan. 20 min. Write about your experience. 10 min.

FARMSTORIES: EDGES

When we first moved to the farm, my parents thought that, if they were clever and could resist buying livestock, we could spend our winters in Florida, in a trailer park by the sea they had found in their travels. And we did that. Each December 19 we would head south on the two-lane highway through the Ozarks, the Appalachians; see barefoot children on porches; eat black-eyed peas and fresh homemade grits; smell the sweet scent of freshly opened orange blossoms.

At school in Florida we learned everything and nothing. Each year we memorized the state capitals, stood nervously by our desks for spelling bees, drew the *S* curve of flamingos. We also learned what it felt like to be chosen last for kick-ball; the names of new birds, new trees; and how to live in and love more than one place at a time.

Read aloud or write and read your own story about bioregions.

# DAY 7

Mind: Brain and Nervous System

Indeed, the ineffability of the air seems akin to the ineffability of awareness itself, and we should not be surprised that many indigenous peoples construe awareness, or "mind," not as a power that resides inside their heads, but rather as a quality that they themselves are inside of, along with the other animals and the plants, the mountains and the clouds.
—David Abram, *The Spell of the Sensuous*

People often confuse the brain with the mind, even though the brain and mind are two distinctly different entities. The mind is "software," the mystical and mysterious product of all that we are. The brain is "hardware," a bodily organ that requires nutrition, rest, use, and proper medical care.
—Dharma Singh Khalsa, M.D., *Brain Longevity*

There is no tissue that is not "body" and no response that is not "mind."
—Deane Juhan, *Job's Body*

The human mind orchestrates our actions and our reactions. If we describe mind as the organized conscious and unconscious mental activities of an organism, we can study the brain and nervous system as tissues that facilitate

Contemplative Shelter for the Ecotarium, Worcester, Massachusetts, by Michael Singer. Photograph by David Stansbury.

mental functioning. We can also consider the ongoing dialogue between the nervous system and the environment, including sense organs that record interactions with light, sound, and electromagnetic fields affecting us moment to moment. In many ways the mind is shaped by its dialogue with place, establishing perceptual and emotional networks in the nervous system. The inner landscape of the mind offers a vast terrain comparable to the outer landscape around us; its study is a contemplation of what it means to be human in relation to the earth.

Many people associate self with patterns of thinking. What is self? Is it brain, heart, soul? Am I my thoughts, emotions, sensations, and intuitions? Do I identify with memories, dreams, imagination, and visions? There are more questions than answers in the study of mind, yet it includes all of the above. Mind has been described as the element or complex of elements in an individual that feels, perceives, thinks, initiates movement or stillness, wills, and especially reasons.[1] Thus, mind develops throughout our lifetimes, reflecting stages of maturity and changes in the environment.

Although the nervous system is complex, each person can have a basic understanding of its components. It is useful to remember that the nervous system interacts with another governing system, the endocrine system, to facilitate and supervise activities of the whole body. The nervous system communicates through electrical signals (impulses) generated by nerve cells and transported along nerve fibers; the endocrine system communicates through chemicals (hormones) transported through the body fluids. As an interconnected unit, the neuroendocrine system helps us unveil the resources and habits of mind that can connect us to or distance us from the world.

Structure and Function

The functional unit of the nervous system is the *nerve cell,* or neuron, characterized by its ability to generate and conduct electrochemical energy forms called nerve impulses. The *central nervous system* (CNS) includes the nerve cells within the brain and spinal cord. The *peripheral nervous system* (PNS) is composed of all other nerve tissue in the body. *Sensory neurons* pick up incoming information and carry it toward the central nervous system; *motor neurons* convey commands away from the central nervous system to muscles and glands so that we can both receive stimuli from the world and act upon them, in constant dialogue with place.

Although the nervous system is essentially one continuous whole, there are further distinctions that can be useful to our study, the *somatic* and *autonomic* systems. The somatic nervous system (SNS) innervates skeletal muscles, causing conscious or voluntary response. The autonomic system (ANS) innervates the smooth muscles of the vital organs, vessels, and glands and is usually considered to be involuntary. The *enteric nervous system* (ENS), which is part of the autonomic nervous system, governs responses within the digestive tract itself. Together, the somatic and autonomic nervous systems provide sensory, integrative, and motor functions throughout

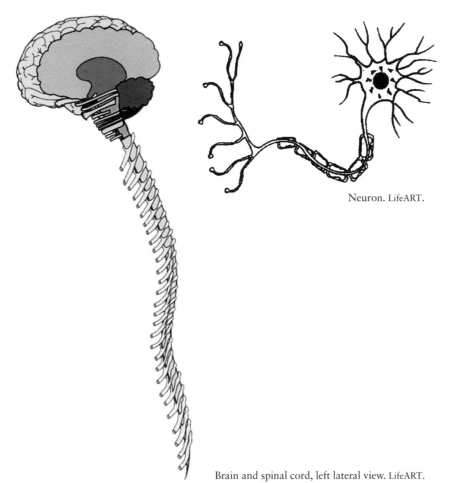

Neuron. LifeART.

Brain and spinal cord, left lateral view. LifeART.

When I was an art major in college, we met for three hours of drawing class, three times a week, working with a model or still life. Day after day, I struggled to capture a dynamic image through charcoal on newsprint, coordinating eye with hand. "Develop a sensual relationship with the model," the instructor called out. "Get inside the experience." But words didn't help; they only escalated my frustration. Eventually, I gave up; released any desire for success. And in that state of open attention, my hand in a single gesture captured the perfect line: my first good drawing. I now refer to this process as the boredom theory of art. For me, boredom is this moment of "dropping through," when the conscious mind is distracted or off guard, and more subtle, integrated connections can occur. Then we engage creative possibility unique to the moment.

the body. Communication between all aspects of the nervous system is fundamental to an integrated experience of the body.

Neurotransmitters are molecules released at the synapses between two adjacent neurons in response to a nerve impulse and at various junctions with other tissues. Neurotransmitters modulate, facilitate, excite, or inhibit neurons, affecting what information reaches brain centers. Although we are born with three times as many neurons as we need, aging or metabolic stress may cause change—decrease or increase in neurotransmitter substances—contributing to memory impairment and motor effects. *Neuropeptides*, another group of chemical messengers (especially prevalent in the digestive tract) signal the brain and viscera, and are part of our internal pharmacy.[2]

Our hominoid ancestors of five million years ago (both apes and humans) probably had brains comparable to those of living apes. Large brains pose a risk for mother and child in birth, as the baby's head must descend through the pelvic hole, but increased size of the female pelvis would restrict efficiency in walking. Thus, to increase brain capacity without further increase in brain size, two characteristics evolved in humans. The cerebral cortex folded to allow more surface area, and the two brain halves lateralized so that specific functions happened in each side. The corpus callosum, a bridge linking brain halves (which first appeared in mammals), expanded to over

Sometimes when I teach dance technique, I become involved in correcting and encouraging students. Their movement loses vitality, originality. As the musician continues playing, I slip out the door and wait for a few minutes, letting them work unseen. When I return, the work invariably has deepened. With less critical attention, exploration fills the room. When we discuss what happened, they recognize the shift. The release from conscious attention offers a doorway to personal work. Creativity, applied through a particular form, teaches us about this dialogue between what we think we are doing and what is waiting to emerge.

two million nerve fibers, allowing rapid, ongoing transfer of information between differentiated brain halves.

The contemporary human forebrain, the *cerebrum,* contains over 100 billion neurons, a comparable figure to the number of stars in the Milky Way galaxy.[3] Weighing about three pounds, the cerebrum uses over 30 percent of the oxygen supply of the body. The surface layer of gray matter, the *cerebral cortex,* is the most highly evolved region of the brain, responsible for coordination of higher nervous activity such as learning, memory, and thought. It is about one-sixth of an inch thick, organized into six layers with at least fifty major subregions. By folding into ridges or convolutions *(gyri),* separated by a shallow groove *(sulcus)* or a deep groove *(fissure),* the cortex increases surface area while maintaining appropriate size. (Only about a third of the cerebral cortex is visible on a brain model!) A *longitudinal fissure* separates the right and left cerebral hemispheres, a *transverse fissure* separates the cerebrum from the cerebellum, and a *lateral fissure* distinguishes the temporal lobe.

Several *sulci* divide each hemisphere into the four prominent lobes, sharing names with the skull bones that cover them: *frontal, parietal, temporal, and occipital* (see diagram). The exception is the *limbic* region, which incorporates parts of the other lobes (frontal, temporal, and parietal) at the base of the cortex. The large right and left *frontal lobes* (forehead) contain the primary motor areas responsible for voluntary movement, in front of the central sulcus. *Broca's area*—a motor speech area for producing language, just in front of the primary motor cortex—coordinates the mouth, tongue, and larynx for speech. Another region governs the refined muscular movements of the hands and fingers for skills such as writing. The frontal lobe facilitates intellectual processes, including reasoning, concentration, planning, abstract thinking, short-term memory, complex problem solving,

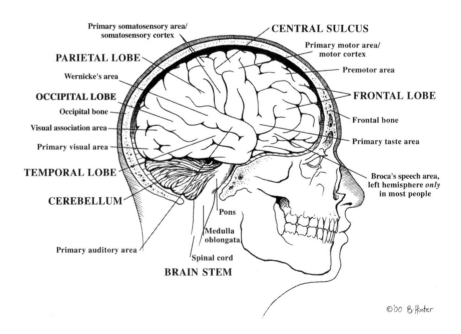

Right lateral view of cerebrum, cerebellum, and brain stem. Anatomical illustration © 2000 by Brian Hunter.

and what we call personality. The *central sulcus* separates the frontal and parietal lobes.

The *parietal lobes* (top right and left sides of the brain) house the primary somesthetic area (*soma*, body; *aisthesis*, perception), often called the *somatosensory cortex*, behind the central sulcus. The somatosensory cortex has a map of the entire body, receiving input from skin, muscles, and viscera (exteroceptors, proprioceptors, and interoceptors) so that you can locate sensation specifically. This region is often drawn as the homunculus, or little man, with enlarged or diminished features, determined not by the size of the body part but by the number of nerve endings present. For example, lips, hands, and feet are exaggerated, due to their large concentrations of nerve endings. Spatially reversed, the feet are on the top of your skull, and the head is on the bottom (above your ears). The parietal lobes also process taste, the use of symbols for language, and abstract reasoning.

The *temporal lobes* (sides of the brain, around each ear) contain the center for hearing, including awareness and discrimination of sound. *Wernicke's area* is specialized for interpreting the content of language. (People with lesions of Wernicke's area can make speech with normal cadence and structure, but it doesn't make sense.) The temporal lobe is also a major site for memory and complex thought processing. The limbic region of the temporal lobe involves emotions such as love, anger, aggression, compassion, compulsion, and sexual behavior. The temporal lobe is distinguished from the frontal and parietal lobes by the lateral fissure.

The *occipital lobes* (the back of the brain) are concerned with vision. This includes receiving, interpreting, and discriminating stimuli received by the eyes. The visual cortex works with other brain centers to analyze visual patterns, like recognizing a friend in a crowd or a deer in a forest.

Association areas are concerned with integrative processing and what we call intelligence, affecting choices about our actions. Association areas give us both the function and the content of the function. For example, the somatosensory association area allows us to understand speech and to read, to recognize body parts and to comprehend their relationships, and to feel the specifics of body and earth as well as to understand interconnectedness. All of the lobes share sensory, motor, and association roles as well as having unique functions. *Sensory areas* interpret sensory impulses, and *motor areas* initiate muscular movement, so that we can feel and respond to the world.

Brain lateralization in humans is a fascinating innovation. To share information and avoid duplication of functions, in most people the left hemisphere of the cortex (right hand) is more developed for symbolic recognition: speech and writing, numerical and scientific skills, sign language and rational, ordered thought. Similarly, for most people, the right hemisphere (left hand) is more important for real-time and space activities and representation of the world: artistic, musical, and visual skills, spatial orientation, and pattern recognition. In cases of injury, some functions can be taken over by the other side. A balance between the two hemispheres offers optimal functioning of the whole body.

When first-graders had difficulty learning to read or to speak aloud, my mother would teach them to skip—forward and backward. This contralateral movement pattern helps integrate right and left brain halves and heightens coordination in the cerebellum. It's also fun, arousing limbic alertness and encouraging emotional integration. Movement affects mind and mind affects movement in a constant exchange. We can begin with either to affect the whole.

Nancy Stark Smith, one of the pioneers of Contact Improvisation, recognizes these subtle connections between mind and movement. Teaching workshops called "Changing States," she encourages dancers to notice that certain kinds of physical activity generate energetic and emotional states and that energetic and feeling states register in movement in an ongoing process. In a teaching situation, if the students are in a gravity-oriented, inwardly focused energy, she can change the overall readiness for a new exercise by altering either the movement or the mood. For example, she can initiate running and jumping, creating levity in the students' bodies—it's hard to be introspective and jump! Or she can use an image such as "feel the air supporting your body" to let the mind become curious about another quality. Throughout students notice the transition between states, the subtle or not so subtle shifts linking one with another.

When I asked a friend his definition of mind, he responded, "The problem came when we made the word a noun instead of a verb. Mind is a process, not a thing."

OF SPECIAL INTEREST: BRAIN MODEL

My niece works for a company called Ma-resh Brainworks, which helps organizations and individuals understand the workings of the brain, based on the triune brain model of Dr. Paul MacLean. In order for concepts to be communicated and understood more easily, the director wears a different hat for each area of the brain. The reptilian brain— the brainstem (a lizard hat)—is about comfort, getting your needs met. If you feel threatened, cold, or have to go the bathroom, for example, you are less likely to be able to attend and understand. Thus, a first step in learning is to meet the basic comfort requirements of the reptilian brain: food, shelter, partnering, and safety. The next hat is the limbic brain (a heart hat), the emotional, relational aspect of the brain, important in memories, smell, hunger, and love. If your basic needs are met in the reptilian brain, you can attend to the emotional richness of interaction, including joy, humor, and all the dynamics of the moment.

The third brain is the neocortex (the thinking cap) for creativity, problem solving, and complex thought processing. The efficient functioning of the third brain is dependent on all the other layers working effectively. For example, if you are mad at someone and your mind is distracted, your decision-making faculties and objectivity will be impaired. The right and left brain lobes balance language and logic (left brain) with intuition and creative problem solving (right brain) linked by the corpus callosum. For efficient learning, both sides need to be engaged. After writing for two years about the complexity of the brain, I liked this simple, clear model, which you can teach in ninety minutes without overwhelming the listener. It makes me want to engage the cerebellum, coordinating movement with all this information, remembering that all parts are essential to the whole (a dancing body hat!).[5]

Both hemispheres participate in sensory and motor functions, memory, and exchanging impulses with other cortical areas. Most motor and sensory tracts cross over from one side of the brain to the other within the brain stem; thus, the motor area of the right cerebral hemisphere generally directs skeletal muscles on the left side of the body, and vice versa. The motor system is asymmetric between the two hemispheres. For example, 90 percent of humans are right-handed.[4] The speech area and the interpretation of language develops fully on only one side, usually the left. (These specializations are not present in our close relatives the chimpanzees.) Thus, the two cerebral hemispheres are structurally mirror images, but functionally they are distinct. Through the bridge of the corpus callosum, it is indeed possible to discuss our next move and create the results simultaneously, drawing on resources from left and right brain halves.

The *diencephalon,* surrounded by but distinct from the cerebral hemispheres, is composed largely of the essential *thalamus* and *hypothalamus.* The thalamus receives all incoming sensory input (except smell) and allows us to perceive sensation generally on the body. The *hypothalamus* governs the neuroendocrine and autonomic systems, integrating emotional reactions with visceral reflexes. Three structures form a diagonal axis through this region. The epithalamus, or *pineal gland,* is the only nonpaired structure in the brain; it responds to dark, monitoring rhythms of sleep and reproduction. The *mammillary bodies,* part of the limbic system, are associated with smell, sucking, and swallowing. The *pituitary gland,* suspended from the hypothalamus by two stalks, responds to light and monitors growth and metabolism. The *limbic system,* the ancient forebrain, spirals around the base of the cortex. This loosely organized group of structures governs emotional/relational aspects of survival and interaction; the *amygdala,* in particular, screens for emotional relevance and monitors the memory aspect of emotion, affecting decisions and attitudes. (See Day 27 Motion and Emotion.)

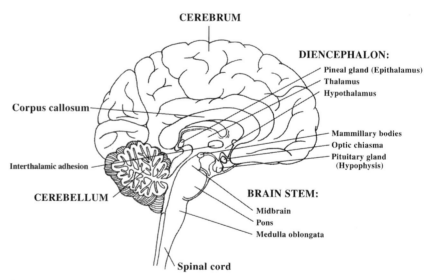

Medial aspect of brain seen in sagittal section.
Anatomical illustration © 2000 by Brian Hunter.

The *cerebellum,* behind and neurologically linked to the upper brain stem, is the second largest structure in the brain. It is responsible for constant movement coordination, maintaining a continuous, cumulative picture of body position in space and storing movement memory. Tracts involving hands and face are particularly extensive for refinement of gestural and facial expression. The cerebellum works with the *basal ganglia,* masses of gray matter deep within each cerebral hemisphere that control large, subconscious movements of skeletal muscles, like swinging the arms in walking. Damage to the basal ganglia results in shaking, rigidity, or involuntary movements. The neurotransmitter dopamine normally inhibits certain motor functions to smooth out skeletal muscle activities; lack of inhibition results in excessive movement, as in Parkinson's disease.

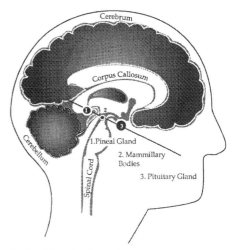

Brain with endocrine axis.

To understand the "lower brain centers" that underlie cortical thought, we begin with the *brain stem,* which connects directly to the spinal cord, and move up toward the cortex. The *medulla oblongata* and accompanying *pons* are responsible for autonomic survival functions, such as breathing, heart rate, and reproduction. The *reticular formation,* with nerve tracts extending from the spinal cord to the cortex, has sensory responsibility for arousal and wakefulness (damage results in coma) and motor functions for maintaining muscle tone throughout the body (readiness to respond).[6] The *midbrain,* a one-inch section connecting the pons to the diencephalon (see above), contains reflexes for focusing attention toward outer stimuli, like turning the head in response to a sound or positioning the eyes. In studying the three major parts of the brain, the cerebrum, cerebellum, and brainstem, we realize that efficient functioning of the "lower" brain centers allows us to think, move, and create with ease.

Brain Health

We nurture our brains; where we live and what we do shapes who we are, whether we are aware of it or not. Thus, we can also look at the development of the brain and nervous system within one human life to understand influences on our thought patterns. The skin and nervous system evolve from the same tissue layer in the embryo, the ectoderm. Thus, touch and movement feed neuronal development. Brain neurons are specified by cortical input early in life, tuning to or developing an affinity for particular stimuli. Each brain cell develops more branches as we age, like a tree, establishing connections or pathways with other nerve cells. The more frequently neuronal pathways are used (like recognizing a familiar person or place), the faster that connection occurs.

The brain grows particularly rapidly in the first two years, as language patterns, emotions, and sensations become linked to thoughts. In the process of childhood development we establish unique circuitry in the nervous system and between brain halves. Large brain cell growth continues into teen years, and we establish more connections than are needed.[7] After that time, we "prune" pathways, giving preference to efficiency and skill in particular

TERMINOLOGY AND FUNCTIONS: BRAIN AND SPINAL CORD

Spinal cord: offers reflexive survival patterns and a two-way communication between the brain and all other body parts.

Cerebellum: integrates and coordinates functions related to movement, balance, and posture.

Brain stem (midbrain, pons, medulla oblongata): a central pattern generator, regulating breath, heart rate, and organ functioning, as well as filtering sensory impulses.

Limbic system: (hypothalamus, thalamus, and midbrain—the parahippocampal and cingulate gyri, hippocampus, dentate gyrus, amygdaloid body, septal nuclei, mammillary bodies of the hypothalamus, anterior nucleus of the thalamus, olfactory bulbs, and bundles of interconnecting myelinated axons): emotional-relational brain, important in memory.

Thalamus: processes all incoming sensory information from the brain and other body parts.

Hypothalamus: governs the autonomic nervous system, which integrates emotional reactions with visceral reflexes.

Paired cerebral hemispheres (outer cerebral cortex—divided into lobes, underlying white matter, gray matter called basal nuclei, and two paired cavities called lateral ventricles filled with cerebral spinal fluid): govern decision making, planning, sensory and motor functions, language, and various aspects of memory.

Corpus callosum: bridge for communication between cerebral hemispheres.

fields of interest and selecting professions, partners, and lifestyles. Brain plasticity and long-term potentiation occur throughout our lives, however. A healthy brain is like a muscle; it needs challenging activity, rest, and proper nutrition for optimal functioning.[8]

Over five million sensory receptors in the body inform the brain, moment by moment, of the dialogue between body and earth. If we are numb to the world around us, we can lose touch with reality, responding to ideas that have nothing to do with actual physical life in the moment. When an event is accompanied by an emotion, such as listening to a song when in love or hearing a disturbing screech of tires, more of the neurological chemicals that create emotions are produced.[9] Later on, we can hear that melody and feel a rush of warmth or hear that sound and experience panic without understanding why. Tension and stress can also affect neurochemical balance, influencing such characteristics as intelligence and personality. Thus, it isn't so much the size of our brains as their neural organization and processing that makes us distinctly human.

Exhaustion of the nervous system is a factor in many illnesses in Western cultures, including autoimmune diseases and anxiety disorders. To relieve neuronal overload we can focus on the sensations of breath, the solidity of bones, and the support of the ground, accessing our resources. We can also walk, resetting the biological rhythm basic to our bipedal structure. Drawing, writing, and speaking invite creativity and emotional expression. As we find a fundamental sense of comfort at the level of the nerves, supported by all the body systems, we can balance the whole. Reflecting on the many dimensions of the nervous system, we expand our potential as caretakers of body, caretakers of earth.

Patterns of mind (alone or with a group)
20 minutes

Seated in postural alignment on a cushion, a chair, or with your back supported against a wall, eyes closed (place a watch or timer near by, set for ten minutes, or designate someone to call time):

• Focus on your breath. Become aware of the subtle sensations of air as it passes through your nose or mouth, nonjudgmentally.

• If thoughts arise as you breathe, just observe them, keeping your attention focused on breath. (You may greet your thoughts, "Hello, thoughts. Right now I am observing my breath. I will attend to you in ten minutes.")

• Your mind may be very busy; it may interrupt you, entice you, and criticize you. Sometimes it helps to simply say "thinking" when a thought comes; name it and let it go. 10 min.

• Even if you can focus on breath for a few seconds without distraction, you have begun the process of awareness, noticing the landscape of mind.

• Pause in open attention. Add vision; write about your experience. 10 min.

You may continue this practice over several days, several weeks, or several years! The task is to notice patterns of mind, without letting them run your life. Psychologists report that we have some sixty thousand thoughts each day, most of which are unimportant, redundant, or even harmful to our health. By observing breath you can be aware of the present moment and have choice about your chattering mind. The moment between perception and response is your moment of choice—to act rather than react from habit patterns of mind.[9]

Photograph by Erik Borg.

When my father developed Alzheimer's disease, I went to visit him in the trailer park by the sea where he and my mother still spend (spent) their winters. He woke me at dawn, took me to see the sun rise up from the ocean, pointed at a freshly opened hibiscus, smiling.

He said to me: "What worries me is that I can no longer get to you in one hop." He said to me: "What I really want is to buy a farm, with a pond and an orchard. We already have all of the animals. But I can't do that. There's not enough time left." He said to me, "See this stone, it's a dead man. Turn it over. See the hole? It's his belly button." He said to me, "Have you finished your schooling yet?" He said to me, "I'm so glad you came, now."

I said to him, "I'm planting apple trees. I'll plant one for you, make applesauce and bring you some at Christmas." I said to him, "Is it scary to forget?" I said to him, "How does it feel to be close to death?" I said to him, "Good-bye."

I said to my friends, "We move too fast for our children and our elders. We need their vision." I said to my friends, "Why do we place so much value on the process of birth and so little on the process of death?"

Read aloud, or write and speak your own story about mind.

Lobes of the brain
10 minutes

Lying in constructive rest, eyes closed:
• Roll the head slowly side to side, stimulating the *corpus callosum,* the bridge between right and left brain hemispheres.
• Now massage the occipital bone (the skull bone touching the floor) with your hands. Imagine touching the right and left *occipital lobes* of the brain behind this bone—the visual cortex of the brain, responsible for vision.
• Move your hands to the areas around your ears (like earphones), the temporal bones. Massage this area and imagine the right and left *temporal lobes* of the brain, responsible for complex thought processing and language.
• Move your hands to the right and left top of the skull, the parietal bones (from a center part, down toward your ears in the center of the skull). Massage this area, the *parietal lobes* of the brain, responsible for a sensory map of the whole body, with the feet at the top of your skull.
• Move your hands to your forehead and massage this area, stimulating the *frontal lobes* of the brain, responsible for short-term memory, planning, and motor commands, initiating such skills as the capacity to speak and walk.
• Move your hands back to the base of your skull, under the occipital lobe, and imagine the right and left *cerebellum* of the brain, responsible for movement memory and coordination.
• Feel the junction between the skull and spine, along the base of the skull—the *brain stem,* linking the cortex to the spinal cord.
• Massage the whole skull and neck. Rock the globe of the skull in a "yes" nod, integrating skull and body. You may feel the connection all the way down the spinal cord to your tail through the dural tube.
• Pause in open attention, breathing fully. Add vision, remaining aware of sensations.

Place visit: Attention to mind

Seated, eyes closed: Focus on breath, noticing patterns of mind. Become aware of the subtle sensations of air as it passes through your nose or mouth, nonjudgmentally. Your mind may be very busy; it may interrupt you, entice you, and criticize you. Even if you can focus on breath for a few seconds without distraction, you have begun the process of mindful awareness. Pause in open attention. How does awareness of mind affect your relationship to place? 20 min. Write about your experience. 10 min.

DAY 8

Visceral Body

The gut has a mind of its own, the enteric nervous system. Just like the larger brain in the head, researchers say, this system sends and receives impulses, records experiences, and responds to emotions.

—Sandra Blakeslee,
"Complex and Hidden Brain in the Gut Makes Cramps, Butterflies and Valium"

The visceral body produces our "gut feelings." When you walk into a room and break out in a cold sweat before you even see who is present, you are experiencing your autonomic nervous system (ANS) at work. The ANS is a primary indicator of the balance of body and environment, connecting vital organs, glands, and skin with the brain stem and hypothalamus. The sympathetic division is responsible for the fight, flight, freeze, or friendly[1] survival response, providing heightened clarity in normal functioning and action in cases of emotional stress and threat. The parasympathetic division encourages rest and recovery, vital to health. The enteric nervous system is the "brain in the gut," functioning largely without input from the brain. Collectively, these visceral layers motivate our actions and reactions, establishing what "feels right and wrong" in moments of choice.

The foundation for experience of place is engagement; we must feel safe enough to walk through the door and interact with the natural world. Comfort precedes expression: A sense of belonging commands the deepening of

Diver—Look Park. Photograph by Bill Arnold.

NEEDS

My colleague Caryn McHose reminds us of the soft support of the gut body and what she calls our bellital alignment. "Plump out your bodies," she encourages. "We sacrifice so much for flatness."

breath and the relaxing of skin, supporting more complex connections. Surprisingly, the meanings of words and descriptions of the visceral body become clearest when one is sitting by a stream on a warm afternoon. Relaxing in this comfortable setting, the structures, emotions, and sensations become one, engaging the participant in an integrated experience. Autonomic and somatic connections merge to shape our embodied perceptions of crystalline sunlight on an easeful afternoon.

Focusing attention on the brain can be foolish—even dangerous—if it is not rooted in autonomic connections. When the visceral body is ignored, we can find our stomachs queasy, moods irritable, heads aching, and vision disoriented. As we return our attention to the needs of the organs and digestive tract through the autonomic nervous system, integration lets us move wholeheartedly through our lives.

Autonomic Nervous System: The Belly Brain

The ANS, also called the visceral nervous system, is composed of nerves and ganglia along the front surface of the spine. Governing the vital organs and glands, the ANS affects heart rate, breath rate, and digestive and sexual functions. The enteric nervous system, a component of the ANS, is found within the digestive tract lining, from mouth to anus. In layered complexity, the human nervous system, including somatic, autonomic, and enteric components, both distinguishes us from and links us to our animal relatives.

ANS *motor nerves* initiate visceral movement and glandular secretions; ANS *sensory nerves* conduct visceral sensations to the spinal cord and brain. Thus, messages go back and forth between body and brain. The hypothalamus of the cortex governs the visceral body and also monitors the endocrine system, integrating emotional reactions with visceral reflexes. Thus, the ANS can function rapidly and continuously without conscious effort, regulating the visceral activities that maintain the body.

The ANS is divided into two parts: the *sympathetic* and *parasympathetic* nervous systems. The sympathetic division stimulates the body toward outer activity and engagement; the parasympathetic, toward cycles of rest and digestion. Sympathetic ganglia (thoracolumbar nerves) run bilaterally from the base of the skull to the coccyx, like a string of white pearls on each side of the spinal column, linked at the bottom as the ganglion of impar.[2] Stimulation of the sympathetic division results in increased activity of the heart and lungs and decreased activity of the digestive tract. When we are about to give a talk or run a race, heart rate speeds up and digestion suspends.

During normal situations of alertness, the sympathetic nervous system supports clarity and directness of action. During what is perceived as a stressful situation, however, sympathetic nerves trigger the fight, flight, freeze, and friendly response: dilation of pupils for increased vision, dilation of capillaries of the lungs for more oxygen, decrease in digestion and salivary gland secretions, increase in blood to skeletal muscles and decrease of blood to the digestive organs, and increase in blood glucose concentration in preparation for activity, as well as release of bladder muscles to reduce energy expended in heating urine.[3]

Concentrated grouping of sympathetic ganglia can be found in the solar plexus and belly. These highly vascularized regions interweave with fascia from the crus of the diaphragm and the iliopsoas, essential for breathing and postural integration. In martial arts and movement training techniques such as tai ji (tai chi) and dance, the autonomic visceral nervous system is considered the location for "centered" energy, called the *hara* in Aikido and the *tant'ien* in tai ji. ANS innervation engages endocrine secretions and autonomic survival responses for speed and efficiency, creating fast, integrated movement.

The parasympathetic division (cranial-sacral nerves) features the vagus nerve (cranial nerve X). From its place of origin in the brain, the vagus nerve meanders the length of the torso—from the throat, down the esophagus and through the diaphragm, along the digestive tract and to the colon—connecting impulses from the hypothalamus and brain stem to all the vital organs and glands along the way. The parasympathetic nerves that emerge from the sacrum (forming the pelvic splanchnic nerves) stimulate muscles of the rectum and bladder and dilate vessels of the reproductive organs (penis and clitoris). The sympathetic and parasympathetic nerves work together, not antagonistically, to coordinate body functioning for optimal health.

The parasympathetic nervous system is activated when the body is ready to relax and digest and there is time for integration, like a quiet afternoon after a big meal. In this situation the heart rate and breathing slow down, and the eyes relax and water. As the digestive system activates, increase in secretions and peristalsis sometimes cause familiar stomach grumbles. Reduction of blood and glucose levels in brain and skeletal muscles causes drowsiness. This restful digestive process allows recovery and integration for all the body systems. Often, it is during the parasympathetic state, including daydreams and doodling, that things start to "make sense," and we have a clear insight or image about seemingly disparate thoughts and experiences.

The vagus nerve, an essential component of parasympathetic activity, is considered a possible link in the development of emotions and sociability in mammals.[4] Its original function in fish was to link the digestive system and vital organs with the evolving brain stem and to conserve energy by slowing heartbeat and digestion in times of reduced oxygen supply. As life got more complex, the vagus nerve retained this function but also developed a new branch to stimulate activity in case of emergency, to speed up systems in case of threat in the familiar fight-or-flight response, increasing heart rate and preparing the body for activity. This includes the freeze response, as in a frog or lizard, rendering oneself invisible by not moving.

With the evolution of mammals, including humans, a third division developed. This vagal nerve complex controls the face and throat, facilitating facial and verbal expressiveness. It also inhibits sympathetic reactions, keeping us calm so that we don't waste precious energy in emergency situations. When threatened or insecure, we can smile, growl, raise our eyebrows, look stern, stick out our tongue, yell, or converse politely as a protective

When I teach about the fight, flight, freeze, or friendly response, I am reminded of this story: I was a junior in college, leaving for a year abroad in Paris. Our group met in Chicago, and we were assigned individual hotel rooms. As I went upstairs, I remember a man in a black suit as one of several passengers on the elevator. When I unlocked my door and entered, he put his black, shiny shoe in the door—I remember it in detail—so that it wouldn't close. Then he forced his way in, locking the door behind him. What surprised me most was my response: I began to talk, launching into a tirade about how it was men like him who make girls like me have to be afraid, and I continued in an irritated and commanding tone. And he left.

In the moment of decision, my autonomic nervous system registered that I couldn't fight (he was stronger), couldn't flee (the door was locked), it was useless to freeze, so I talked. This "friendly" response—our capacity to smile, make facial gestures, talk, or otherwise verbally persuade—is a survival mechanism that many of us use all day long.

A busy physical therapist talked with me about her clients. "Even though I see many preoperative and postoperative patients," she said, "I treat most people for exhaustion of the autonomic nervous system. To begin to heal, they simply need me to touch them, to help them calm down and rest."

When he was sixteen, my stepson wrote us from a wilderness canoe trip in Quebec, describing the usual rigorous portages and mud fights. Then he added: "I get lots of free time here. It's great." Our contemporary lives are so busy, even our children find free time an adventure.

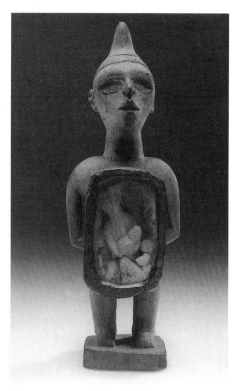

Power figure, lower Congo area, southwest coast of Africa. Wereldmuseum, Rotterdam, The Netherlands. Photograph by Erik Hesmerg, Sneek.

mechanism.[5] In other words, talking and acting friendly can be a form of autonomic protection. Thus, we can use words to communicate, or we can converse to distance ourselves from involved interaction.

A cyclic balance between silence and speaking, rest and activity, is fundamental to the healthy functioning of our internal organs (including the brain). We all know the situation, however, of eating a big meal and running off to a meeting or a physical workout. In this case the nervous system sends contradictory messages: the parasympathetic nervous system is activated to stimulate the digestive system for integration, and the sympathetic nervous system is activated to deal with high-level functioning in the world. One physical result of these conflicting messages is indigestion. If this imbalance is a constant occurrence, the further result is nervous system exhaustion as the various components function inefficiently and are denied time for rest and recuperation.

For many of us, it is difficult to establish a situation that allows enough time, trust, and relaxation to activate the impulse for rest. Individuals who are under constant stress or who have been "on guard" for prolonged periods (such as war veterans, students during exams, parents tending young children, or children of abusive parents) often find it hard to regain a balance between the sympathetic and parasympathetic nervous systems. Part of the process of healing the nervous system is to establish an environment in which we can let down our guard and be supported, without the pressures of alertness for survival. For many, being in nature stimulates a sense of well-being, allowing profound integration and recovery. As the sympathetic nervous system relaxes its guard, the parasympathetic has time for rest and integration. Increased blood flow to the vital organs encourages us to yawn, stretch out on the grass, and nourish our core.

Enteric Nervous System: The Brain in the Gut

The enteric nervous system, or the "brain in the gut," is composed of a primitive nervous system called the neural net within the lining of the digestive tract. As we have seen, the enteric nervous system is part of the peripheral nervous system (all neuronal tissue outside the brain and spinal cord) and of the autonomic nervous system. Yet the brain in the gut has one unique characteristic: it can function autonomously, processing sensory information and maintaining homeostasis, with little or no interaction with the brain and spinal cord. Formed by over 100 million neurons (more than the spinal cord), the enteric nervous system involves two layers of tissue, the myenteric plexus and the submucosal plexus, surrounding the esophagus, stomach, small intestines, and colon. Studies show that it can act independently, learn, remember, and feel.[6]

Each of us can recall situations where we followed our instincts or gut feelings. We made a phone call at the right moment, helped a friend, or met someone at an unexpected moment. Functioning below rational thought, our visceral body picks up on information that may not register in our conscious minds. This "instinctive" aspect engages another kind of intelligence and knowing, balancing our picture of ourselves as brainy and smart. Various models have been offered to explain the seemingly mysterious process of unconscious knowing, such as biologist Rupert Sheldrake's morphogenetic fields or psychologist Carl Jung's descriptions of the collective unconscious.[7] Yet in our daily lives we can notice that, when we engage all the dimensions of our nervous system potential, we feel supported by, rather than in conflict with, our deepest motivations.

⬚   ⬚   ⬚

Pouring the gut body*
5 minutes

Standing, or seated comfortably:
• Place your hands on your chest and lean your body forward, pouring the organs of your lungs and heart toward your hands until you feel some of their weight.
• Pour your weight back to center by returning to vertical alignment.
• Place your hands on the center of your body, over your waist area. Lean forward and pour your weight into your hands. The stomach, liver, pancreas, large intestine are falling toward your hands as you change their relationship to gravity. Try to let them relax and respond to the change in support. Return to vertical, pouring the weight back toward center.
• Place your hand on the belly of your body, over the small intestines and reproductive organs. Lean forward and let the gut body fall toward your hands for support. Pour back toward center.
• Try walking or dancing with your gut body; notice how it feels to keep this area of your torso soft and responsive.

The box†
30 minutes

• Imagine a box; let it be a size you can carry in your hands. Be specific: how does it look, feel, smell? Think of what is most present in your mind at this moment—something that returns again and again, wanting your attention. Put this in your box, close the lid, and place the box somewhere safe, where you can easily find it again when you are ready. 5 min.
• Do the other exercises on these pages: pouring the gut body and inner observer.
• When you have finished, go to your box, open the lid, and notice what you have put inside. Has it changed?
• Write about your experience and what you put it the box. 15 min.
    Read aloud to yourself or to a group. Allow your voice to stay connected to your "gut" feelings. 10 min.

*Title and exercise from Caryn McHose, movement teacher and educational bodyworker.
†Title and exercise from Doug Anderson, poet and teacher of creative writing.

FARMSTORIES: THE CARDBOARD BOX

The cardboard box was found under the attic eaves of our old farmhouse after the move. *(breath)* It wasn't found by the first owners: the ones who cut down the cherry tree and felled the outbuildings *(breath)* so they could plow the dark, rich, soil all the way to the back door. *(breath)* It was found by the second owners: the ones who tore down the giant barn—in fine shape too, the neighbors said—*(breath)* so they could make their own mark on the land.

    The cardboard box now sat on the dining room table at my sister's house, a few miles down the road from our old farm. We were gathered there for my father's eightieth birthday, and the contents of the box were to prove beyond a doubt something I had never quite believed in all of my years of growing up: that my parents had ever loved in quite this way—before they were married, beyond themselves.

(continued on p. 54)

The box was filled with black-and-white photographs of my parents' courtship, the wedding, with new babes, standing on the land. There was one photo each of my mother and my father, taken the morning after their first overnight together on the sailboat my father had built in his grandmother's basement: my mother holding a cup of coffee, leaning against the mast; my father, at the tiller, staring off into the distance. They were each beautiful and handsome in turn, serious, in trouble. It was one of those moments when you see your life passing in front of your eyes, and you watch its unfolding. So when my mother's parents objected to their marriage—a different class, an artist to boot—they were in trouble, but they kept on going. They kept on going.

The cardboard box was placed under the attic eaves when we moved to the farm in 1952, and it was abandoned, but not discarded, in the move back to the city in 1961. This was the same move in which my mother's trunk, filled with white dresses and linens, christening gowns, wedding attire, had been unconscionably sold to whoever would buy—an act that brought my mother to tears, then to rage, then to despair, for months and years and decades after.

Why don't I remember the move from the farm? Don't open the cardboard box until you're ready.

Read aloud to yourself or to the group, or write and read your own story about the visceral body. Maintain awareness of breath. Pause at the end of each sentence or phrase and take a full breath—inhalation and exhalation. Allow a final breath cycle at the end of the story.

Inner observer*
15 minutes

Seated or lying on the floor, eyes closed:
• Imagine a nonjudgmental and supportive inner figure inhabiting your body.
• Now allow this observer to travel slowly from head to toe, through each part of your body, noticing whatever is occurring.
• As you move your awareness, notice colors, sounds, shapes, and smells. Notice sounds, words, or images. Notice bones, organs, and blood. What draws your attention? What's going on for you today?
• Imagine how this inner character would speak about the experience.
• Open your eyes, remaining aware of sensation.
• Invite a word or words to come to you to describe your experience. Be simple, noticing how language emerges from experience.
• Speak this word or phrase out loud: "Cold." "Peaceful." "Trembling like leaves."
• Invoke your inner observer at several different times of the day. This nonjudgmental and supportive aspect is another part of yourself, one you can easily access at any moment, equal to your usual identity.

Place visit: Attention to the visceral body

Receive what is happening around you without feeling that you have to do anything. Feel the soft front surface of your body in a receptive, digestive state. Be sure the jaw is relaxed; allow air to pass in and out through your mouth as well as your nose. What do you notice? Walk around your place with your gut body relaxed. How does it feel? You may feel nothing, or have an urge toward action; notice that. 20 min. Write about your experience. 10 min.

*Title and exercise drawn from Alton Wasson of the Contemplative Dance Workshop.

✳ DAY 9

Perception

We all have sense organs which are similar, but our perceptions are totally unique. . . .
Perception is about relationship—to ourselves, others, the Earth and the universe.
—Bonnie Bainbridge Cohen, *Sensing, Feeling and Action*

We construct our view of the world through our senses. Billions of receptors throughout our structure constantly feed us signals about ourselves and about our surroundings. Our ability to organize and interpret these signals is called perception. Our response, which may be to not respond, completes the process. Throughout, we are active participants: we can heighten awareness of sensing, broaden patterns of interpretation, and encourage new pathways of response.[1] Understanding this perceptual process can help us *act* from the sensory information available at the moment, rather than *react* from habit and outdated association, enhancing our ability to respond—our responsibility. Ultimately, what we perceive determines what we know—and what we think is real.

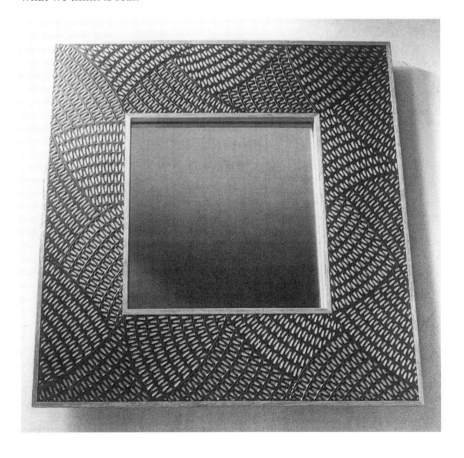

Mirror, by Kristina Madsen. Bubinga, ebony, dried pearwood. Photograph by David Stansbury.

We construct our ideas about reality through the specificity of human perception. The nervous system responds to only a selective range of wavelengths, vibrations and other stimuli. For example, human eyes utilize a particular spectrum of light for vision; it is the same spectrum of light that plants use for photosynthesis, but it is not the full spectrum. A bee employs light in the ultraviolet range for navigation—an aspect of light that our eyes don't recognize. Human ears are equipped to respond to a particular range of vibrational frequencies. These are the same frequencies employed by birds, which may partially account for human fondness for their songs, but it's not the full range of sound. Whales and bats use higher frequencies. There is more, not less, to life on earth than we perceive.

Perception is also culture-bound, shaped by education, religious and political systems, language, and arts. For example, instrumental music that might evoke a trance state in Morocco could sound simply irritating or chaotic to an uninitiated Western listener. American educational systems are generally oriented toward visual-auditory learning. Literacy in the areas of movement, touch, and visceral response (including emotion) is often ignored, limiting our capacity for subtlety of expression, communication, and problem solving. Senses respond to particular landscapes, smells, tactile stimulation, gestures, and sounds; our interpretation of this input becomes our experience, based largely on cultural conditioning.

Perception is also personally selective, affected by genes, family values, previous experience, and the current state of alertness. Beyond very basic, inborn survival patterns, we learn to focus our senses. No two people perceive the same reality. Ask a group of thirty engaged in the same activity what they are noticing, and each will have a very different view. In a conversation, unique and sometimes contradictory interpretations will occur. We each have perceptual strengths and biases. For example, one person may choose from visual input, another from auditory or tactile stimuli.

Expectations affect perception: we see what we expect to see. For example, if you are looking for your friend at an airport, thinking that she has red hair, you might miss her if she has dyed her hair black. In some cases, broadening expectation may allow us to detect subtler signals. When we look at each other, most of us see solid bodies. But if we learn that electromagnetic fields exist around our forms and are measurable in a laboratory, we might experience the body more expansively. We are used to feeling a person's blood pulse; but if we learn that there is also a craniosacral pulse, slower than the heartbeat (6–12 cycles, instead of the 60 to 70 heartbeats per minute), we might then begin to perceive the craniosacral rhythm. Our expectations about perception can limit or expand our perceptual range.

In this three-part model, which includes the human perceptual system, the cultural, and the personal, our perceptual range becomes progressively more limited. However, through information and experiential exercises, we can reinhabit our fuller potential. Perception itself is highly plastic. As we begin to recognize and change the filters of association and categorization through which incoming signals are screened, our perceptual conditioning and habits can expand.

EXPECTATIONS

I was trained to see. My father was a painter. When we went on family vacation, the children would pile out of the car, stand silently, and look. I have numerous photographs of my sisters and me lined up on a rock, staring into the distance. I thought this was how all families perceived the world. When I traveled by car with my friend, her siblings jumped out, ran here and there, climbed all over everything, exploring. They had to move through the landscape to know it. Another friend, whose family members were musicians, listened to the sounds around them, sang, and talked, inhabiting place through sound. Each person, family, country, and culture has priorities for perception; we get into trouble when we think ours is the only way.

There are four generally accepted "categories" of sensory receptors: the *interoceptors,* found mainly in the organs, are responsible for monitoring the inner workings of the body, such as blood chemistry, heartbeat, and digestion. (If you've just had coffee, you may feel a "caffeine high"; if you need food, a blood sugar "low.") The *exteroceptors,* found in the skin and connective tissue, are responsible for monitoring the outer environment through "touch," including several kinds of sensations such as pressure, heat, cold, pain, and vibration. (You might notice the light touch of cloth brushing against your skin, the pressure of your feet against the floor, or your pelvis onto the chair.)

The *proprioceptors,* found in the joints, ligaments and tendons, muscles, and the inner ear, are cumulatively responsible for registering movement, balance, and body position in space. (If you shift how you're sitting, your proprioceptors tell you that you are moving and the resulting position.) The *special senses* include sight, sound, smell, taste, and touch. "Touch" is used to refer to all the exteroceptors in the skin. The proprioceptors and interoceptors are generally omitted from our education about perception. Yet our experience is not limited to the outside of our bodies. It is essential that we include the signals provided by proprioceptors and interoceptors in our perceptual awareness.

Smell (olfaction) is the oldest of our special senses, responding (in humans) to chemicals in the air. Olfactory hairs filter air moving through the nasal passageways in the nose. Olfactory glands (at the hair bases) both keep the muscosal lining clean and help dissolve unwanted chemicals. Registering directly in the olfactory bulb of the brain (part of the limbic system), smell is closely linked to survival responses. Often interpreted below conscious awareness, smell affects emotional memory, sexuality, and "gut feelings." Taste receptors are also chemoreceptors. Located in the taste buds (fungiform papillae) of the tongue and (to a lesser degree) the soft palate, these receptors register bitter, sour, salt, and sweet flavors in various combinations. The taste of the foods we eat (or reject) is strongly influenced by smell, and also by texture, temperature, visual appeal, and personal history.

Perception is cumulative, generally simultaneous, and necessarily selective. Rarely does one part of the sensory system act alone. Nearly all sensory signals go first to a relay station in the thalamus, a central structure in the brain, for integration with other sensory input and then to primary sensory areas of the cortex and "higher centers" for interpretation and response. For example, dimensional interpretation of a visual image is cued by information from touch, smell, taste, sound, sight, movement, and visceral activity. In the big view, all senses are one. Some aspects of sensing occur more rapidly than others; smell registers almost immediately, affecting relationship to the foods we eat, the places we visit, and the company we keep.

Perceptual selectivity is essential. If we were to be aware of all of the sensory information coming into our bodies at any moment, we would be overwhelmed. Thus, where we place our attention (which is a motor activity involving the proprioceptors) affects what we perceive. For example, you might be drawn to the sound of the telephone ringing while focusing on

Once I drove from New England to California with a Yale architecture graduate, eager to share an architectural tour of the country. We began at Frank Lloyd Wright's Fallingwater *in Pennsylvania, passed on through Louis Sullivan's towers in Chicago, journeyed to Wright's* Bartlesville Tower *in Oklahoma, visited Paolo Soleri's experiment in the Arizona desert at* Arcosanti, *stood on the resilient wooden floors in Arata Isozaki's spacious art museum in Los Angeles, rested in Michael Graves's light-filled library in San Juan Capistrano, and arrived in San Francisco to take up our own creative work. What was fascinating about the journey was our varying perspectives. I would look at a field and see glorious space; he would see a building site. I would experience a structure from its interior space; he explored the exterior. I would think of implied content (bodies moving); he would think of form. It taught us both about our assumptions and our edges.*

Pre-Columbian gold pendant, fifth–tenth
centuries, Colombia. The Metropolitan Museum of
Art, The Michael C. Rockefeller Memorial Collection,
Bequest of Nelson A. Rockefeller, 1979 (1979.206.497).

reading. The nervous system prioritizes change, which could mean threat or opportunity. What was background suddenly becomes foreground. Intention affects attention, and both are influenced by our general state of health, alertness, and motivation. Sensory information is also selected at perceptual gateways that necessarily inhibit and delete large quantities of stimuli and determine which goes on to other brain centers for interpretation. In neurological language, we have to open the attentional gates for stimuli to pass through and be made conscious, and the most frequently used pathways are most easily accessed. As we become aware of perceptual habits, we can make a practice of inviting new information, opening new opportunities for response.

Projection also affects perception. Some psychologists consider that as much as 90 percent of what we perceive "out there" is actually "in here." Projection is the tendency to place our own feelings, desires, and fears outside ourselves and assign them to other people, to place, and to objects. Often it is a seldom noticed aspect of ourselves, projected into the environment. For example, if you see a stranger and decide she is angry, you can be projecting your anger onto that person, without acknowledging that it is possibly partly internal. If someone is walking toward you and you think he or she is being critical of your hat, you are likely projecting your own critical views outward.

Projection also occurs on a cultural level, where "other" is construed as a threat. Projecting our own unexplored nature outward poses "difference" as an excuse for violence, war, and environmental destruction. In both a personal and a cultural context, unacknowledged tendencies toward projecting inner experience outward are limiting and potentially dangerous. Yet projection is part of perception; our experience of the world around us is filtered through our inner life. Often, if we are cluttered internally, we experience the world as cluttered. If we are irritable, we recognize irritability. As we become aware of the process of projection, we can engage its useful dimensions. At best, we see outside ourselves what is inside, enhancing consciousness. Evoked in theater or dance, for example, projection encourages us to experience our own potential within the full range of human experience. Recognizing projection as an aspect of perception, we become aware of our potent role in constructing our view of the world.

Numbing is also a phenomenon in our era. We are bombarded by sensory stimuli, much of which has emotional content beyond our capacity for response. Our daily dose of television news and images about the escalating environmental crisis can require that we protect ourselves by shutting down sensory pathways. For example, as we look at a color photograph in *National Geographic* showing a Norwegian beach strewn with the dead babies of puffins (from the lack of fish in those waters) it hurts. We don't want to see. Although "shutting down" is a healthy, even lifesaving response in specific situations, it is limiting, even dangerous as a constant state.

Environmentalist and author Thomas Berry calls us the "autistic generation." We don't see, we don't hear; nothing gets in, nothing gets out. As we become less receptive, we miss the delight, sensuality, and depth of feeling that comes from being engaged with life. However, once we can recognize a continuum of perceptual receptivity from fully open to fully closed, we can choose the degree to which we want to open to the environment around us and the degree to which we need to protect ourselves for the moment, creating temporary but useful boundaries. At the level of the cell, the selectively permeable membrane monitors what enters and leaves, moment by moment. Receptivity changes, allowing us to participate actively in the process of sensing, feeling, and action.

As we recognize that perceptual habits have a great deal to do with how we interact with the world around us, we can engage our human perceptual system for maximum responsiveness and health. Perception is the basis for

When my husband spoke of getting homing pigeons, I resisted. Then he told me of their extraordinary sensory capabilities. Author Stephen Bodio's delightful book, Aloft, *details their abilities: they use the sun as a compass and detect the faint electromagnetic field of the earth. They notice local variations in gravity and can discern changes in barometric pressure, such as an approaching storm. They hear infrasound, like wind over a mountain range or oceans at a distance of hundreds or even thousands of miles; they see polarized light and colors within the ultraviolet range. Pigeons can even smell their way home when prevailing winds are constant. Studies at Cornell University's Laboratory of Ornithology show that their sensory palette is both diverse and cumulative: if one area is limited, others make up the difference. We now have sixteen pigeons.*[2]

connection. Acknowledging possibilities and limitations reveals to us our responsibility for conscious interaction with the world. The more refined our awareness of sensory input and interpretation, the more choice we have about response. As we increase perceptual range, we can invite heightened connection to self, to the sensual pleasures, and to the landscape around us: the dialogue between body and earth.

◻   ◻   ◻

TO DO

Naming the sensory receptors
10 minutes

Lying in constructive rest or seated, eyes closed:
• Notice a place on your skin where you feel light touch—perhaps cloth or air brushing against your arm. Then notice deep pressure, like the weight of your foot pressing into the floor. Notice any areas where you feel pain, heat, or cold. Notice vibration, like the sound of a voice. These specific feelings are sensed by your *exteroceptors,* with specialized nerve endings in the skin.
• Attend to your organs, notice sensations of digestion, blood pulse, or blood chemistry (low blood sugar, too much caffeine). These inner workings of your body are monitored by the *interoceptors,* in your organs and blood vessels.
• Begin to roll or stretch, following any impulse for movement that feels good. How do you know what new positions you are in? Move again, noticing body position. You are stimulating your *proprioceptors,* with unique nerve endings in muscles, tendons, joints, and the inner ear.
• Gradually add your *special senses.* Notice sound (the touch of sound waves vibrating your eardrums and body tissues), smell (the chemicals in the air registering through receptors in your nose), and taste (the chemicals affecting taste buds on your tongue). Open your eyes and add vision (light waves touching receptor cells in your retinas).
• Notice the connection between the inner landscape of your body and the outer landscape around you.

Seeing and being seen (the witness)
20 minutes

Seated or standing in a comfortable position, eyes open:
• Choose something, living or other-than-living, on which you can focus your attention for several minutes: a tree, a pond, a chair. Find a comfortable position and allow yourself to witness, nonjudgmentally, for five minutes. If your mind wanders, bring it back to the process of conscious witnessing, observing whatever occurs.
• Pause in open attention. Now allow yourself to be witnessed by the "thing" that you chose. Imagine you are being witnessed by your tree, pond, or chair, nonjudgmentally, in whatever you are doing. 5 min.
• Pause in open attention. Write about your experience; read aloud to yourself or someone else. Witness as you listen. 10 min.

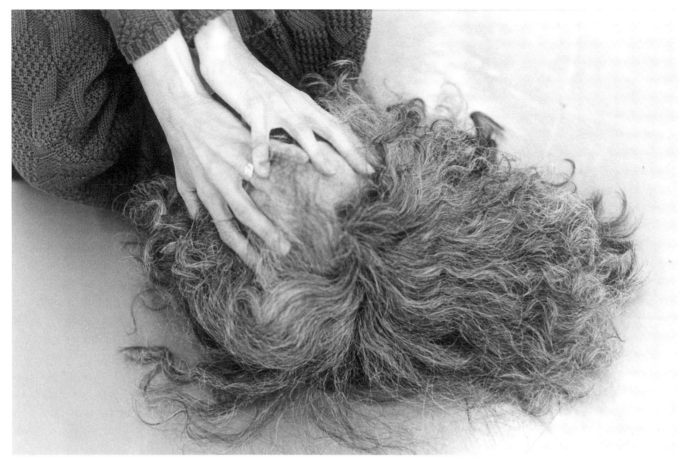

Authentic Movement project.
Photograph by Erik Borg.

Place visit: Expectation and projection*

Seated in open attention, explore how expectation and projection affect your awareness. Choose two aspects of your place within your field of vision, like a tree and a stone. (If in a group, sit in a circle around them.) Begin speaking out loud, one at a time, using the words "I expect that you . . ." For example: "I expect that you would be rough to touch. I expect you will stay where you are." Continue until you have voiced many expectations. Then change to "I am the one who . . ." For example: "I am the one who protects you from sun and rain; I am the one who is very hard." Continue, nonjudgmentally, until you have voiced (and listened to) many projections. 20 min. Write about your experience. 10 min.

*Exercise developed from Susan Harper's work with "Em'oceans and Sensations."

FARMSTORIES: SKY

When I was a child on the farm, we would watch the clouds to know our fortune. Weather is life to a farmer.

My father, a watercolor painter, taught me to recognize the signs by naming the colors. Payne's gray could mean the summer rains to crack open the hard shells of the seeds, start them growing. Too much gray could keep the farmers out of the fields, rot the roots of the corn. Cobalt blue could be the endless sunny days of summer to nurture the crops; a cloudless blue could parch the earth, crack open the soil, wither the leaves of soybeans. A bit of alizarin crimson along the horizon could warn of the hailstorms that would dent the hoods of trucks, shred the wheat. Hunter green mixed with the clouds could foretell the tornadoes that would rip paths through the fields, tear the roofs off barns.

As a child, I would lie my back on the ground in the fields and watch the sky to know our fortune. It always changed.

Read aloud, or write and read your own story about perception.

✳ DAY 10

Touch

To know through touch . . . is to understand better . . . the ways in which lines of communication between human beings and all other forms of organic and inorganic life can remain open if the currents that run through the hand are encouraged to flow.
—Michael Brenson, "Memory of the Hand," *Sculpture Magazine*

As we touch, we are touched. When we touch a tree, we also feel the tree's bark touching us. We can be aware of the world outside us or the world inside us. In this way, touching connects us directly to place but also to ourselves. Memories, thoughts, and imaginings flood through our touch, creating what is termed subjective experience: the information perceived is as much about the toucher and the toucher's history, as it is about what he or she is touching. Yet touch also provides immediate physical objective information, basic to our survival, discerning weight, size, texture, and temperature.

Body, tactile sculpture by Rosalyn Driscoll. Steel, wood, leather, epoxy resin. Photograph by David Stansbury.

STROKING

Brittany spaniels are hard to manage. When we brought our puppy home, I could hold him in the palm of my hand and calm his fear through touch. As Tobie grew, this brown and white ball of energy became the emotional barometer of our household. He telegraphed when someone was leaving, the prevailing mood, and any moment of excitement. Sometimes, I would look at him, to know what I was feeling. "Bold but sensitive," the bird dog trainer wrote on his report. When someone approached our house, he would run in circles, do flips off the door. No command would deter him. Eventually, I learned to put my hand on his chest and stroke, letting my energy meet his. If I can calm myself, he can respond.

Touch and movement underlie all other senses. In the developing embryo the various receptors involved in touch are already active as early as week 7, before those of hearing and vision have formed.[1] As infants, we literally "make sense of" our environment through touch and movement, exploring body, space, and objects as a single and vast tactile playground. Then, in a process developmental movement analysts call "measuring," we differentiate self from other by first experiencing the distance between hand and mouth, then by reaching out for an object, then by pushing away from or crawling toward a parent, then by walking through a doorway. Gradually, we develop a sense of space, and the time it takes to move through it.

Language reflects the degree to which touch and movement are linked to our feelings of well-being. We use the terms "touched" or "moved" when deeply affected; we are "out of touch," "removed," when distant from others or ourselves. We describe difficult situations in body terms: if someone is "on our backs," or a "pain in the neck," we long for release. The word *sense*

Hands on tactile sculpture, *Elegy,* by Rosalyn Driscoll. Steel, wood, marble. Photograph by Rosalyn Driscoll.

itself implies reality—to make sense. Yet language can draw us away from or toward our senses, abstracting the moment or engaging the uniqueness of place, describing or transmitting experience.[2]

Skin is the primary organ of touch. This semipermeable membrane covers the entire body, spanning sixteen to twenty square feet and maintaining both a barrier to and a connection with the environment. The skin of an adult's body weighs about eight pounds, with a thickness varying from a few epithelial cells on the eyelids or behind the ears to ten or more tough layers on the soles of our feet. The epidermis is the outer, air-exposed layer of skin. Composed of stratified squamous cells without blood vessels, this tissue must continuously replace itself as dry cells slough off. The dermis, or inner layer of the skin, is thicker and richly interwoven with sensory receptors and nerves. Blood and lymph vessels, fibrous connective tissue, smooth muscles, sweat glands, and hair follicles are also found in the dermis. Connective tissue is one continuous sheath throughout the body, wrapping every tissue, including muscles, joints, vessels, and the insides and outsides of bones as it moves toward the core. Thus, connective tissue fibers offer strength and elasticity to the skin, linking outer sensory receptors (exteroceptors) with deeper layers.

Touch receptors are present in every part of our skin.[3] These exteroceptors (outside receivers) include specialized nerve endings for light touch, pressure, vibration, cold, heat, and pain. Some receptors, such as free nerve endings, are on the surface of the skin; others, such as deep pressure receptors, are below the dermis, extending into the deeper tissues of the body. Specific nerve ending receptors include *Ruffini's end organs* for deep continuous pressure and joint compression; *Pacinian corpuscles*, registering deep pressure and vibration; *Merkel's disks, Meissner's corpuscles*, and *hair end organs* for light touch; *Krause's corpuscles* for cold; *Ruffini's corpuscles* for heat; and *free nerve endings* for pain and light touch. Of course, there may be exteroceptors that have not yet been identified.

The proprioceptors (self receivers) discern movement and location of body parts in space. These specialized nerve endings can be found in the skeletal muscles, the tendons in and around joints, and the inner ear. *Muscle spindles* tell us about muscle length; *Golgi tendon organs* detect muscle force and the pull on tendons; *joint receptors* monitor compression in our joints; and *maculae and cristae*, sensory hair receptors within the three fluid-filled semicircular canals in the inner ear, detect equilibrium.

Interoceptors (internal receivers) are free or encapsulated nerve endings in the walls of blood vessels and viscera that inform us about the inner workings of body, including *chemoreceptors* (chemicals), *baroreceptors* (pressure), and *nociceptors* (pain). They merge this information with the special senses—the *photoreceptors* of the eyes (light); *taste buds* of the tongue, soft palate, and epiglottis (chemicals); *olfactory hairs* in the nose (chemicals); the other *touch receptors* in the skin; and *auditory receptor cells* in the ears (vibration)—to give us our full perceptual matrix.

Skin serves diverse functions, both protective and communicative. It responds to inner and outer stimuli, changing size and shape and permeability.

Sitting at the computer, hour after hour, muscles transport mind through touch. John Updike, speaking of writing, says, "Our task as we sit (or stand or lie) is to rise above the setting, with its comforts and distractions, into a relationship with our ideal reader, who wishes from us nothing but the fruit of our best instincts, most honest inklings, and firmest persuasions." All this is recorded in the sensitive stroke of finger to key.

Many of the impulses to the brain for movement are necessarily inhibited each day, allowing efficient functioning in society. Psychologist Jean Houston humorously describes her remedy for relieving such tight constriction by getting in an elevator alone and doing all the movements she couldn't express during the day, relieving the nervous system. Many of us can empathize with the sensation of sitting in meetings and having to "hold still" for too many hours. Our bodies are longing to move and to shake out excess tension—to tremble, stretch, or wriggle in discovery and recovery.

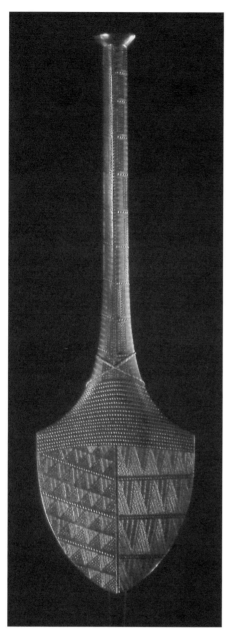

Kinikini club (Fijian paddle club), by Makiti Koto. Yesi wood, 40 in. × 12 in.

Skin is also involved in temperature regulation, resistance to microorganisms or puncture, toxin release, vitamin D synthesis, gaseous exchange, blood pressure regulation, storage of food and water, and lubrication. In its living, breathing exchange with the environment, skin informs our concept of what exists outside and what exists inside the body, shaping our identity and our capacity for distinction of self and other. Skin is intimately related to the nervous system, in that it originates from the same embryonic layer and the same primitive cells. Rashes, hives, and blisters offer visible evidence of nervous system stress or toxins in the body, calling for our attention. Touch is a way of directly communicating with the nervous system, calming us or alarming us.

The hand is a specialized organ of touch, comparable to the ear, eye, nose, and tongue. In the evolutionary history of mammals, our primate lineage developed hands and feet instead of paws, hooves, or flippers. Ape ancestors brachiated in treetops and hunkered on branches, challenging the evolving nervous system with complex demands for neuromuscular coordination. As the representation of the arm and hand in the brain (and brain size in general) increased, natural selection favored changes suitable for this new, multidimensional way of life. Eyes moved forward for distinct focus, and the thumb gravitated opposite the fingers for ease of grasping and articulation, while the fingers folded squarely to palm. Tactile pads on the ends of fingers and toes were richly supplied with touch receptors; hairless palms and soles with abundant sweat glands ensured a moist and responsive grip.[4]

Increased articulation of the hand shaped and reshaped the developing brain. The human hand wrapped around things with the thumb and the two outer fingers folding over the palm, creating a spiral with the wrist and forearm. By the time of *Homo sapiens,* the refined hand was an extension of a dimensional arm, linked to a lateralized brain, giving distinct priority to processing information from each hand while coordinating the pair. As we touch a tool created by the hand for the hand, like a Passamaquoddy stone gouge made for shaping bowls or a carpenter's wooden plane "handed down" through the generations, we feel its elegant subtlety. The brain forms and is informed by the objects around us.

The lateralization of brain function gives each hand its role. Rather than thinking of dominance and subordination of hand or brain hemisphere, consider the collaboration and cooperation enabled by this structural development. If we throw with our right arm, the left counterbalances the weight; if we chop vegetables with our left, we hold them with our right. Both feed their sensory input into the body as a whole, forming and informing our perceptions and choices. The corpus callosum is the bridge between the two halves of the brain, creating the essential dialogue for integration and collaboration. So important are the hands to human life that a specialized subregion of the cortex evolved specifically to oversee their functioning.

Communication can be observed in hands, whether in touch or in gesture, sometimes in direct contrast to words. There are various kinds of touch expressing the full range of human emotion and thought. From a feather-light brush to an enraged and violent action, our bodies and hands carry out the impulses of mind. Intention is expressed through the body. In "conscious touch," in which we are aware of self and other, connection is both receptive and expressive. Our personal experience of touch may be clouded or uncertain, yet we can rehabilitate its role, reaffirming its importance in our lives.

Touch is essential to physical and emotional health, a fact repeatedly documented in studies on both animals and humans. Statistics in orphanages show that touch deprivation affects physical and emotional development and can result in retardation or death. Premature babies in hospitals develop 50 percent faster if given tactile stimulation. Research conclusively links violent crime with deprivation of physical pleasure and harsh childhoods, including trauma at birth. Still, with all the statistics available, we are a culture starved for touch and inhibited in movement.[5]

There are frequent situations where touch and movement receptors shut off or turn down. Injury, abuse, and repeated strain or stress can numb our bodies. Critical attitudes or judgmental images of body can distort perception. Each of us has places of decreased sensitivity, forgotten or unexplored landscapes. We also have favored areas with heightened awareness. Still, our bodies are highly plastic and responsive, and there are many ways to restimulate or balance sensitivity. Three basic steps help us begin: change attitude—respect body intelligence; change action—feed supportive touch and movement into body systems; and allow dialogue—listen when the body communicates its needs.

□　　□　　□

Layers of touch

5 minutes

Seated, eyes open:
• Place your hand on your thigh. Rub your thigh lightly, focusing on the texture of cloth covering this part of your upper leg.
• Continue rubbing, but bring your attention to the skin under the cloth. Skin is a smooth, semipermeable membrane; be aware of sliding your fingers over this layer below the cloth as you rub.
• Bring your attention to the muscles under the skin. Thigh muscles are stringy, with parallel fibers (like steak or chicken); sometimes they feel hard as rock; but as you continue to rub, they will soften, responding to the heat of your touch.
• Slide your hand down to the back of your knee and feel for the ropes or cords: the tendons, which attach muscle to bone, and the ligaments, which attach bone to bone. Rub these dense, parallel, collagenous fibers of connective tissue that weave the joints together; note their texture.
• Now feel for the bone, the densest tissue in the body. It is easy to feel bone at the knee, but you can also feel it in the thigh, under the layers of muscle. Remember that bone is living tissue; it produces red and white blood cells in

A captain of our football team took my anatomy class and reflected on his attitudes about touch: "I remember the exact day it was no longer OK to touch my friends," he said, "but it was OK to hit them and to hit them hard." Now he was studying to be a doctor; he needed a new relationship to touch.

TO DO

When we moved to the farm, my mother's friends encouraged her to find a project to fill the long winters' nights. She chose a white quilt, with pale blue thread to embroider the houses and birds and barns that made up the design. It took nine years. I remember her seated on the living room couch, the quilt spread over her lap, fingers darting up and down, connecting thread to thread, while we danced or sang or fought around her.

When I had my own home, my mother gave us quilts to fill the empty rooms. There was the farm quilt. A red and white one, offered right off my parents' bed when admired on a visit. A navy blue and yellow one— boy's colors. And lap quilts, one each, for our drive across the United States.

At first, I thought the quilts too pretty, fearing their House Beautiful, picture-perfect look in my own home. I folded each one carefully, placed them in drawers. But now she calls on the phone, telling me that her fingers are too numb to hold a needle, that my father sets the table for five and asks where we are.

And when I finish talking, I wrap each quilt around me one by one, feel their warmth, let them hold me.

Read aloud, or write and speak your own story about touch.

the body and changes with use, abuse, and nutrition throughout your life. Become aware of the sensation of bone as you rub.

• Reverse the process, starting with bone. Continue rubbing and feel for the ligaments or tendons at the back of the knee. Feel for muscles. Feel for skin. Feel the cloth covering the skin.

• Now remove your hand from the cloth and notice how long you can feel heat radiating between your hand and thigh. Notice when this feeling goes away.

• Place your hands on the floor to clearly separate the connection.

Basket of objects (alone or with a group)
20 minutes

• Gather several objects that are interesting to you, and place them in a basket (or have someone else do the collecting). Consider gathering natural objects from the area where you live, small enough to carry in your hands.

• Close your eyes, and select one object through touch. Hold it in your hands and feel it carefully. Bring your attention to the texture of what you are touching, to its temperature and density. Take your time to connect with it through image or imagination.

• Open your eyes slowly and look at your object. Notice what changes by adding vision. Then, allow looking to invite more touching. 5 min.

• Close your eyes again, and explore through touch. As you stay with the process of touching, where does your mind take you? What associations come, beyond naming? Bring your vision back; remain aware of the sense of touch as you look. 5 min.

• Write about your object and your experience. Read what you have written aloud to yourself or a group. 10 min.

Place visit: Attention to touch

At your place, eyes closed: Begin touching—the bark of the trees, the soil, rocks. Remember that as you touch, you are being touched. Try touching with different parts of your body: your hand, the side of your face, your lips, bare feet, and your back. There are receptors for light touch, deep pressure, temperature, vibration, and pain in all parts of your body. Add vision to your experience of touch and movement; notice that the motor impulses to change attention are also movement! 20 min. Write about your experience. 10 min.

Photograph by Erik Borg.

☀ DAY 11

Movement

We all have reasons for moving. I move to keep things whole.
—Mark Strand, "Keeping Things Whole," in *Reasons for Moving*

Rain (1989), choreography and performance by Bebe Miller. Photograph by Martha Swope.

Movement partners the dynamics of the earth. Momentum, centrifugal force, and gravity inform the way we swing an arm, turn our head, or leap across a stream. In many ways, the moving body is our connection to the family of things. Rather than creating movement, we participate in the inherent movement of the planet. Physicality encourages an embodied experience of place. You can't, for example, slide to the ground and be other than where you are; motion requires your immediate attention. Yet intention is sometimes unclear; as we seek critical praise and approval, we can begin to consider a strong or flexible body as the goal, rather than the process through which we connect to the larger forces of life.

We can trace a growing bias against the intelligence of the body in Western cultures for the past two thousand years. Gradual institutionalization of religion, medicine, and education has left physical intelligence and creativity suspect. Trust in our own inner processes, in particular, has been discouraged. This distancing of ourselves from ourselves is in part responsible for the split of self from other and of humans from the larger dance of the cosmos. It is as though, by denying awareness of our earthly form (the moving body), we avoid "evil." And that theory has proven disastrously inadequate. Instead, it is time to revitalize our awareness of what it means to be human in all our dimensions and to make choices accordingly. To engage the challenges of the environmental crisis, we need our full selves present—our physical, emotional, intuitive, thinking, sensual, and spiritual selves.

Movement occurs in the context of planet Earth. Our bodies are formed and informed through our relationship to this unique environment of gravity, geomagnetic fields, radiation, and air pressure. We push our weight down into the earth to take a step forward; we lever against the ground in order to reach or turn around. (Try doing these movements on slick ice!) The gesture of opening a door engages muscles throughout the body— some to stabilize, some to mobilize. Movements register the qualitative tone of an experience: whether we are anxious, joyful, tentative, or bold; whether a place is threatening, pleasurable, mysterious, or enlivening. Where there is life, there is movement.

Neurologist Oliver Sacks, in his book *A Leg to Stand On,* considers kinesthesia (sense of movement), our sixth sense, essential to an understanding of self and the world. Movement is a pervasive, ever present part of our lives; cessation of movement in the nervous system is a sign of paralysis or death. Generally, we move our bodies in order to touch. As nerve signals stimulated by the environment travel from the periphery to the core of the body (spinal cord and brain) and motor impulses respond, they create an ongoing feedback and feed-forward process of exchange. Even when we place a hand on someone and remain still, our intention moves through the layers of tissue, registering skin, muscle, nerve, connective tissue, blood, and bone in the other body and ours.

In a simplified view of a conscious motor activity, the brain sequences movement for efficient interaction with the environment. Premotor pools in the spinal cord and brain stem create stereotyped and patterned response. (In dogs and cats these are "hard wired," creating predictable behavior.) The primary motor cortex *produces* the movements to accomplish a goal. (In this process, muscles work collectively to carry out a specific function or command, rather than individually.) The premotor cortex *plans* movements. (With a lesion in the premotor cortex, you can't put together a planned sequence of movements.) And the prefrontal cortex *sequences the programs* that produce the action. The cerebellum, working in conjunction with the basal ganglia, coordinates the whole, feeding back to every part of the cortex through the thalamus, the primary relay station for all sensory input. Anything that damages this web of connections,

DANCING IN PLACE

When I first did tai ji (t'ai chi) outdoors in a public park in China, I was not alone. Older and younger bodies moved silently around me, in and out of tree shadows. I was struck by their ease in moving as part of the landscape. Now, when I walk outside and lift my arms overhead, extend a leg into the air, or rotate my spine to encompass the globe of space around me, I remember that it is possible for movement to appear natural on the earth. When I create a performance outdoors, my goal is for dance to seem normal rather than odd, amid the trees and clouds and pathways of our lives.

A student said it took her four years to become oriented to living at Middlebury College—the region and the place. As a Native American who had participated in pow-wows since childhood, she danced imagining herself on the Nebraska plains, on the reservation. Her junior year she traveled to India and France and returned determined to spend her senior year "dancing in place." She began each studio visit by orienting herself toward home, then honoring the directions: north, south, east, west, up, and down—until she felt herself in the center of her own energy. Then she could dance where she was, fully embodied.

interrupts efficient motor functioning. Pathways in the nervous system that have been used before for interpretation of sensory input and for motor coordination are the most readily accessible, like a trail that's been well used. These habits in movement and thought can be useful and/or limiting. Just as with the landscape, as we move off the familiar roadways, new dimensions unfold.

The Place of Dance

Although one may say, "I'm not a dancer," in reality each of us participates in the dance of existence. We choreograph our days, pattern our movements, ritualize our spatial pathways, and communicate through subtle and not so subtle bodily indications. More specifically, we each develop a relationship with our body. We breathe, touch, rotate, stretch, flex, and shape the ever changing events of the time. We also encounter transpersonal connections—moments of grace or heightened awareness that are beyond conscious intent. Whether or not we choose to amplify this dialogue through African dance classes, partnered swing dancing, or private living-room improvisations, we are part of a larger dance. Humans move and participate in movement around us; it is our basic nature.

There are many experiential models for exploring this vast terrain. Within the discipline of Authentic Movement, for example, one can experience movements from one's own personal history (like stretching, attending to injuries, or indyosyncratic movements), collective movements (like synchronous gestures, unplanned encounters), and transpersonal experiences (becoming a vessel for voices and images beyond personal experience). Janet Adler has described four different sources that she has studied within her practice of the discipline of Authentic Movement: the collective unconscious, the personal unconscious, personal consciousness, and collective consciousness. "Transpersonal experiences can arise from any of these sources; however only when they appear in relationship to personal or collective consciousness can they become integrated into the psyche and thus in support of the health and well-being of individual and collective life."[1] These sources are ways of describing different aspects of our movement life and reflect layers of possible awareness. As we replace fear of our moving bodies with focused attention, we can learn how rich our inner movement world really is.

Movement forms have long been used for health of individuals and of communities, for healing and recovery, to access spiritual dimensions, and as artistic expression, indexing the arc of human life from birth to death. We each need a strong body and a well-developed knowledge of self (a mature ego) to function as a vessel for the larger energies that move through our physical forms without being overwhelmed. For this, we keep our minds adaptable and integrated with our moving self; we practice articulating the spine in all three directions every day; we open and tone our abdomen; and we allow our breath to be full and available. We also invite our eyes to be mobile rather than fixed, necks to be pliable rather than rigid, hands articulate rather than grasping and controlling, skin relaxed and permeable rather than tight, fluids active rather than stagnant, and our base of support—our feet—sensitive to the earth.

In our mobile culture, it also helps to consciously identify aspects of spatial orientation: north, south, east, west, up, and down. Sometimes it is useful to orient ourselves to "home," wherever that might be, by facing in that direction before beginning movement. Walking, particularly at a pace faster than we can think (two steps per second), integrates and enlivens the nervous system, wakes us up, and brings us to our senses. Disorientation is useful as well. When we are fatigued by our daily decisions, overinvolved in certain areas of our life, actions such as rolling, moving backward, spinning, and falling can invite lower brain integration, bringing release (and relief) into other realms of experience. As we literally lose our cortical "grasp" on reality, the dialogue between control and surrender is an evocative theme.

Movement is fundamentally about rhythm, one of the basic underlying connections of body and earth. Rhythm is when things happen, occurring

Once I received a grant to make a dance a day as I traveled across the United States. At some point, I would get out of the car, walk onto the land, and move, spending five minutes or an hour with the terrain. Removing all need to choreograph or perform, movement took its own form. Following impulses, the rhythms, gestures, and motions became a conversation between my body and the landscape of the moment.

Indian, Shiva as Lord of Dance, bronze, c. 1000, ht.: 69.4cm. Kate S. Buckingham Collection, 1965.1130. © 2000, The Art Institute of Chicago, All Rights Reserved.

When I arrived on Paros, Greece, to teach a workshop, my eyes were blinded by the glistening white buildings against the Aegean Sea. What space could I use for my studio? I wondered. In that landscape it seemed contradictory to remain inside to dance. Soon I was escorted by Greek dancer Vasilis to the countryside on his moped, cassette tape player in hand. We danced that afternoon amid the olive groves, near an old monastery, goats and birds as our audience. Movement was playful; it was serious; it was the core of our lives.

on various time scales, from the repetitive sound of a heart beating moment by moment to the seemingly isolated eruptions of volcanoes over billions of years. Rhythm is also a way of establishing community, including human and other-than-human participants. When we move together in rhythmic interaction, we experience relationship; when we are aware of the overriding rhythms of the seasons, epochs, and eons, we feel ourselves as participants in an expansive choreography. Historian and critic Arlene Croce refers to rhythm in dance as poetic suggestion. "That is the level on which dancing, by the alchemy of its rhythm, transfigures life."[2]

Although movement holds us to present time and place, it is also rooted in the past. From the origins of biological life, movement reflects the impulse to condense in or expand out, to go toward or away, to open or to close. All the physical laws of the planet are inherent in our bodies. Movements disconnected from underlying physical integrity are those that create injury or loss of vitality. As we begin to learn about our specific bioregions, we can enhance connection to the unique topography of each landscape and the underpinnings of cultural movement forms. Moving with awareness links past with present and with the future.

From a cultural perspective, we have been preceded on this American continent by native peoples with a ten-thousand-year relationship to the land, including a heritage of songs and dances for every occasion, every season. Yet many of us are recent arrivals to this continent, with less than a four-hundred-year history, incomplete stories and often emotional ambivalence about this place we call home. We also have brought with us our heritage of other landscapes, and the songs and dances, language patterns, and movement motifs they have inspired—whether we are aware of them or not. The task for many over the past century has been to seek a new relationship with our bodies and with this new land, to deepen connections between body and earth.[3]

We can build time for movement into our lives by expanding familiar patterns of daily activity. Get out of the car and walk through the landscape; wear clothing that facilitates movement and feels good on the skin. Try walking on new streets and visiting unusual places, like a botanical garden or a park, that stimulate the senses and evoke different kinds of awareness. Cook a meal outdoors, camp, hike, watch birds, go fishing, or just meander together. Leave wallets at home; enjoy what is right there, free and inviting.

Research confirms what we have known all along: learning skills of the hand, skills of the body, at a young age develops neurological pathways used throughout our lives. Learning to dance, to play music, games, and sports can enhance our children's lives as well as our own. Build rituals together: take a hike once a month or a walk each evening. Include a star-gazing moment before bed each night. Allow time to unfold in relation to the rhythms of the earth, rather than remaining indoors all day. Notice the difference in health, friendships, and family.

Yielding and standing (alone or with a group)

10 minutes

Standing in postural alignment, eyes open:
- Begin walking. As you move through and around the space, feel the texture of the floor under your feet, the air on your skin and inside your body, sounds and smells; notice boundaries, walls, or objects, other people or things.
- When you are ready, continue walking, but imagine (or experience) a hand stroking lightly down your back, signaling release of your spine and hip joints.
- "Pull the plug" and let yourself yield to the ground; soften your body and find support with your arms and muscles as you need so that you don't bump your knees. Connect your movement with the stroke of the hand so that touch and falling occur simultaneously.
- When you reach "zero energy," roll and return to standing and walking.

With a group: Everyone begins walking. Without speaking, stroke your hand down someone's back, or allow yourself to be stroked. When you feel the contact, both partners begin to yield to the floor, timing the fall with the movement of the hand. Find zero energy, roll apart, and begin walking again. Stroke or allow yourself to be stroked as you walk.

Authentic Movement (the mover)

20 minutes

Lying on the floor, eyes closed:
- Follow any impulses for movement that feel good, as when you first wake up in the morning and move before you think about moving.
- Scan through your body as you move to be sure all parts are included, with no hierarchies or aversions. Movement may be tiny or big, as feels appropriate to the moment. Open your eyes if you feel like moving fast or through space.
- You may find yourself "being moved" as you move. The body knows exactly what it needs at any moment for recovery, rest, or expression. Authentic movement is a process of body listening, following deep inner impulses of the body, nonjudgmentally.
- Remember, there is no right, no wrong. Take this time to listen; let the body tell you its stories. As you continue, movements that want your conscious attention return again and again until they are recognized. There is no need to "grasp" or try to remember.
- When you are ready, pause in open attention.
- Add vision; take a few minutes to identify three moments that you can remember from your movement for any reason: an image, a specific shape, an energetic state, or a gesture. This is like taking a big marking pen and drawing a circle around a moment, clipping an image for a scrapbook.

Through the discipline of Authentic Movement, the mover develops a dialogue with the rich movement language of the body.[4]

OF SPECIAL INTEREST:
THE DISCIPLINE OF
AUTHENTIC MOVEMENT

For over twenty years, I have participated in a movement practice that is now called the discipline of Authentic Movement, by Janet Adler. The ground form, as Adler calls it, appears to be simple, yet a complex relationship evolves between a mover and a witness. The mover closes his/her eyes and follows impulses in the body for movement or stillness. The witness, with eyes open, intends to bring awareness to his/her judgments, projections, and interpretations, in service of a desire to be only present. This process develops to include many movers working in relationship to a circle of witnesses. Such foundation has been enriched by the work of Mary Whitehouse and Janet Adler regarding what Adler calls the development of mover consciousness within the discipline of Authentic Movement.

The experience, for the movers, initially involves slowing down so that all levels of movement are available, relaxing outer vision so that we can register internal states. In all the years of working in this way, I remain amazed at the connections that occur—dynamic back arches and head stands beyond my conscious capabilities, exquisite partnering, and transcendent states, like the voice of a Chinese woman calling to me across a mountain stream. As always, experiences like these are hard to describe, impossible to measure; yet they heighten our awareness of all that has meaning. Sometimes, after a session, conversing in a local pub, I imagine everyone in the room shifting into an authentic state— following the inner impulses of their bodies, acknowledging connections that we normally ignore.

FARMSTORIES: MODELS 2— THE BOAT

The boat model was made by my father as a child. Carved of wood and painted green, it had three tall masts, intricate rigging, and a captain's wheel. The boat model traveled with us from house to house, not as a prized possession but as one of those objects that have inspired too much care to be casually tossed away. It was a model of concentration, of patience, of a young man wanting someplace to put his attention and his dreams.

Perhaps the boat model was made while his mother, my grandmother, carved the wooden puppets for her W.P.A.-funded puppet shows, which she gave during the Depression. Or while she wrote horoscopes for the local paper and romance novels in the afternoon, absorbed in her own life. While his grandmother, my great-grandmother, prepared the food and fed the boarders who paid the rent. Imagination was essential.

The boat set a course in my father's mind that only he could know. The important thing, perhaps, was that he saw himself as captain of his own ship, and that we as children, when we saw his boat, knew we had our own ships to make and to sail.

Read aloud or write and read your own story about movement. Orient yourself within the specific place where you are speaking. Notice light, air, objects and people as you read.

Dance a day
One week

• Make a dance a day, each day for a week. Move for five minutes, maintaining intuitive connections to yourself and to the specifics of place; follow impulses as they come, rather than constructing a sequence to remember.
• You may want to do your dance a day in outdoor, private settings so that you can fully explore movement; or you may want to do "invisible dances," simply moving as you normally would move in a specific time and place (in your kitchen, driving to work, talking to a friend) but considering it a dance. Be clear about your intention when you are doing your dance a day.
• After each five-minute dance, take time to write about your experience. Notice how place affects your movement.

Place visit: Attention to movement

As you walk to your place, focus on the physicality of your movement (including your hands). When you arrive, begin with orientation: face each of the four directions, connect up to the sky, down to the earth, and end by facing toward your home. Then, seated or standing, eyes closed (to start), with awareness of breath: notice *sensations*. Pause, in open attention, and begin noticing impulses for *movement*—the urge to stretch your spine or leg, to move your neck. Bring your attention to these impulses for movement and allow yourself to follow. Consider sensations as the language of the body, inviting expression. Close your eyes as you move or wait for movement impulses; continue scanning your body, asking what it would like to do rather than thinking up movements. Follow whatever feels good. After twenty minutes, write or draw from your experience. 10 min.

DAY 12

Sound and Hearing

The bristly bundle of "stereocilia" at the top of the cell quivered to the high-pitched tones of violins, swayed to the rumblings of kettle drums, and bowed and recoiled, like tiny trees in a hurricane, to the blasts of rock-and-roll.
—Jeff Goldberg, "The Quivering Bundles That Let Us Hear,"
in *Seeing, Hearing and Smelling the World*

Sound touches us. When an object of any size vibrates, it sets in motion the particles of the air, creating a wave pattern of alternating bands of high and low pressure. Hearing is a conversation with these sound waves: a three-way interaction between vibrational frequencies in the environment, sensory receptors in our inner ears, and neurochemical signals in the brain. Sound waves also vibrate all the tissues of the body, including our bones, muscles, organs, and glands. We receive sounds continuously as our ears remain attentive day and night, vulnerable to noise. We also create sounds as our feet vibrate the earth, our voices and cars echo into the air, affecting the lives of

Honey in the Ivy. Photograph by Bill Arnold.

Film scores, I am told, rely on sounds of which we are only peripherally aware (not just the music track) to convey emotional tone, set the stage, and build anticipation. Sounds register whether we are conscious of them or not. Dogs barking, an unexplained click or shuffle, children's voices, or a door opening inform our sense of safety or alertness at an unconscious level; in fact, background sounds are more effective dramatically if they remain under the conscious threshold. Radio announcers also play with our leap to interpretations. On Garrison Keillor's Prairie Home Companion, *various characters and sound effects, including Fred Farrel's animal calls, remind us of all that can be visualized from a sound.*

other living organisms. Amid the myriad sounds that accompany our daily lives, we focus on only a few with any clarity—those that engage our attention. Yet we are affected by all the vibrations in the environment, whether we are conscious of them or not.

In the process of hearing, sound waves are funneled by the *external ear* into the *external auditory meatus (canal)* of each ear. As the sound waves touch the tiny sheet of *tympanic membrane* or eardrum, the membrane vibrates, converting sound energy into the mechanical energy of touch and movement. These vibrations travel through three tiny bones, or ossicles: the *malleus* strikes the *incus,* which strikes the *stapes* (known as the hammer, anvil, and stirrup because of their shapes). The stapes, the smallest bone in the body, transmits the energy to the flexible *oval window,* a watertight membrane between the middle ear and the fluid-filled inner ear.

As the stapes jiggles the oval window, it creates wavelike motions in the fluid of the *cochlea,* the sense organ for hearing. (Within the temporal bone, the snail-shaped, pea-sized cochlea forms one portion of the inner ear labyrinth, with the vestibule and semicircular canals for equilibrium.) The fluid vibrates the *basilar membrane,* a gauzy strip of tissue just over an inch long. Coiled within the cochlea, the basilar membrane is lined with over sixteen thousand specialized hair cells, arranged in four long, parallel columns. Hair cells respond more readily to different frequencies at different locations along the membrane (high-frequency sound waves at the base of the cochlea, where the membrane is narrow and stiff, and low-frequency sounds at the top, where the membrane is wider and flexible). Bundles of *stereocilia,* arranged like a cone on top of each hair cell, quiver or "dance" with the mechanical vibration of sound waves. This activates the auditory nerve fibers running from the base of the hair cells, changing mechanical vibrations to electrochemical signals.[1]

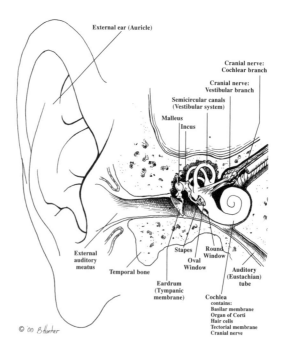

Auditory and vestibular systems. Anatomical illustration © 2000 by Brian Hunter.

The *auditory nerve* from each ear travels to the brain stem and to other relay stations, including the thalamus, the primary relay station for sensory input. If the sound has threatening or emotional content, processing goes directly from the thalamus to the amygdala of the limbic system for immediate response. It also continues to the auditory cortex at the top of the right and left temporal lobes for basic processing of loudness, pitch, and timbre, with hearing from each ear processed in both sides of the brain for spatial location. At this point the nerve impulses just register as sound or noise. For interpretation and meaning, the nerves must combine their data with those of other cortical regions.

According to the complexity of the sound and the degree of engagement, connections are made to various brain lobes. These include the parietal lobe, for association with other sounds heard in the past; the temporal lobe, for interpretation of language and music; the occipital lobe, for visual cues; the frontal lobe, for specific information and refined interpretation, with signals going back and forth to the auditory cortex. As in touch and vision, patterns are detected to determine order amid all the sensory data. Perceiving sound is a highly subjective process, informed by the health of our tissues and our personal and cultural histories.

In evolutionary history, human ears began as gill slits in ancient fish. They were used for filter feeding as seawater passed in through the fish's mouth and out the gill slits. In mammal-like reptiles the first gill slits formed the middle ear cavity, an example of evolutionary efficiency in which former structures are renovated for new purposes. In reptiles this cavity housed only the stapes, but all mammals have three ear bones, a redesign of other structures of the reptilian mandible and jaw to form the incus and malleus. The eustachian tube, connecting the throat to the middle ear cavity, is also from the first gill slit; the eardrum evolved from the second gill slit but moved inside the skull, and the outer ear formed a unique funnel for airwaves.[2]

The mobile neck muscles of reptilian ancestors evolved into mammalian face and scalp muscles. This musculature allowed mammals to move their ears, enhancing hearing. The ability of some mammals to move and rotate ears separately to catch sounds is visible today as a moose stands with one ear oriented forward, as you paddle slowly by in your canoe, and one back toward the woods. Primates relied more on vision and a three-dimensionally mobile spine to monitor their environments, so ears have less movement capacity, although some humans can still wiggle their ears! Adults with average hearing detect sound waves with frequencies of from 20 to 20,000 or more vibrations each second; the greatest range of sensitivity is between 2,000 and 3,000 (the pitch of a child or a woman's cry).[3] Human hearing is less acute than our mammalian relatives, such as cats and whales, which can hear 50,000 and 200,000 cycles per second, respectively.

Sound locates us in space. It also extends our awareness of space beyond our visual fields. Echolocation, the reverberation of sounds off objects in the environment (both by an outer stimulus, like a car honk, and those we make ourselves, such as voice or footstep) creates an awareness of place. Sounds

Loons have an eerie and fascinating call. During June in New Hampshire, I listen to their mix of hoots, wails, tremolos, and nocturnal choruses, plus each male's distinctive yodel. The "wail" is used for communication between a pair with newly hatched babe riding on one parent's back, after a month of incubation. Flying at speeds of up to one hundred miles an hour when they migrate from the ocean back to familiar lakes and mates each spring, they use the tremolo as their only call in flight. Loon researcher Judith McIntyre suggests that even the naming of this ten-million-year-old species of diving bird reflects their sounds: they are referred to as huart *(for "cry") in French-speaking Canada,* gagara *in Russian (imitating the Arctic loon), and* kwe-moo *in Native American languages of the northeastern United States. And in English, "loony" refers to someone odd or insane. Their haunting sounds, varied and rich, are like nothing else, and myths and stories abound.*[4]

A South Siberian Turkic people, the Tuvans use mimesis, or imitation, for aesthetic purposes in the creation of their music. Theodore Levin of Dartmouth College states, *"For the Tuvans, one of the purposes of music seems to be to offer detailed and concrete descriptions of topography—a little like the famous 'song lines' of the Australian aborigines."* Listening to their sonic maps of physical landscapes, explored through the text of their songs (throat-singing, whistling, and other abstract sounds), I felt connected to the natural world. Both the mountains and grasslands of their homeland and unique ways of inhabiting their bodies were amplified through sound.

tell us how near or far we are from walls, trees, cars, and other people as well as the volume of the particular space we inhabit. People who are blind may develop echolocation with particular acuity and also use the reverberation of the clicks of a cane against objects for detailed information about the environment. Background sounds, those not consciously heard, constantly convey a sense of safety or discomfort as well as the specifics of place.

Hearing is an important survival mechanism. The limbic brain registers change. Driving down the road at night, the events of the day might be in the foreground of our thoughts, accompanied by the sound of the tires on pavement, windshield wipers and rain, and our own breath as background. Yet if the tire develops an odd rhythm, the wiper hums or falters, or breath varies, these background sounds instantly become foreground. Depending on the tone of the sound, the brain decides instantaneously how much adrenaline to distribute for the situation, preparing for action. Our ears remain open at all times of day and night, and hearing mechanisms need rest. Sometimes in our noise-laden society, we can wear ourselves out simply by sound pollution, keeping the body constantly on alert without adequate time for recovery.

Moving, hearing, and sound making are intertwined in humans since birth, beginning with the first breath. Movement continues to underlie hearing throughout our lives. We turn our heads and bodies to position our stereoscopic ears for optimal hearing. The inner ears function both for hearing and for balance, informing our dynamic relationship with the earth, with our own voices, and with the moving body. Sound moves us as well. Sound waves travel through and vibrate all the tissues of our bodies. Low-frequency sounds, with long waves, resonate in larger chambers of our structures, like the pelvis; high-frequency ones, with shorter waves, resonate in smaller cavities, like sinuses in the skull. Different densities of tissue respond to different frequencies. We are constantly being vibrated by the environment around us.

Hearing can be enhanced and developed. We know that our capacity for hearing develops and changes throughout our lives. Studies show that, for children, musical training in particular helps strengthen neural connections and thicken the corpus callosum, linking right and left brain halves.[5] Hearing involves silence as well as sound. Visiting quiet places revives our capacity for careful listening. We can develop listening skills by focusing our attention on the detailed sounds in the environment, learning to recognize the "who cooks for you" song of the barred owl or the drumming of grouse in early spring. If we use our ears, they remain responsive and alert.

Hearing loss also occurs. Most of us fail to realize how delicate and how easily and irreversibly damaged are the hair cells in our ears. There are only 16,000 hair cells in a human cochlea, compared to 100 million photoreceptors in the retina of the eye, and they are extremely vulnerable. Decibels are units measuring sound intensity; their count starts at zero, the intensity of the least audible sound to a normal human ear. A whisper is about 40 decibels; conversation, 60–70; heavy traffic, 80; a rock concert, around 120 (which may cause discomfort); and a jet plane takeoff, 140 (which causes pain and possible hearing loss). Sound over 90 decibels is physiologically damaging and can cause deafness. Standing too near a fire alarm on a routine drill can damage ears for life. Hearing loss also can occur through age, as the eardrum thickens and loses its capacity to register high-pitched or soft sounds.[6]

In many ways, we learn to let information come in one ear and go out the other, because we cannot cope with the quantity or the quality of stimuli. Selective hearing is part of our lives. With our remarkable ability to discern a familiar child's cry on a playground full of children, the voice of a loved one across a crowded party, or an uncertain tone across telephone wires, our hearing is multifaceted and unique. To actively engage our hearing we need to offer opportunities for silence as well as sound, rest as well as activity, and safe listening environments as well as the fast-paced classrooms, workplaces, and streets, which can overwhelm our sensory receptors.

Authentic Movement project. Photograph by Erik Borg.

Language

When a baby is born, all sounds and all languages are possible. The process of language making develops in a specific time frame, beginning with various communications, through crying for the first weeks (gradually differentiating pleasure, pain, and hunger) and continuing during the first year. Concepts attach to specific sounds, like recognizing that the sounds "ma-ma," in English-speaking countries, means "mother." At around six months the muscles of the mouth, lips, and tongue are developed enough to create vocalizations imitating speech, and cultural patterns determine specific language tones and rhythms. Studies show that babies' hearing begins to favor sounds of their native language—that spoken by the parents—by four months, excluding other sounds at neurological gateways.[7]

Language is an abstraction of sound, requiring processing in several regions of the brain. In other primates, vocalization reflects emotional states such as alarm, fear, or affection, and the limbic ring is the primary area that initiates the calls. Researchers have found that three distinct sounds in a vervet monkey, for example, are used to warn of leopard, python, and eagle.[8] But in humans, language is more complex, requiring its own processing centers.

Limbic stimulation still evokes shrieks, laughter, and emotion-laden sounds beyond the control of the cortical mind. Try stifling a scream on a roller-coaster ride or an exclamation when startled, and notice the speed of limbic-oriented vocal response. Thus, sound and hearing, vocalizing and speaking, involve a rich dimensional conversation with the world around us.

Vibrations*
5 minutes

Lying on your belly, forehead lifted slightly off the floor, eyes closed or open:
• Create a closed bowl by cupping your mouth, one ear, and the floor with your hands.
• Make the open sound, "ah," and sense the vibrations in your face. Continue, and notice vibrations in your head, arms, and torso.
• Make the closed sounds of "ng," "n," "mm" and sense those vibrations.
• Play with other sounds. Where in your body can you move the vibrations? What sounds vibrate the most? The least? Exaggerate the consonants. Exaggerate the vowels.
• Tell a story, perhaps what happened this morning, continuing to sense vibrations.
• Pause in silence; then proceed. Rest in open attention.

Audible breath
5 minutes

Seated or lying, eyes closed:
• Place your hands on the bottom of your ribs. Make the sound of "ah" with your breath. Notice how the diaphragm moves as you breathe. It attaches around the bottom of your ribs and to the front of your spine, like a horizontal trampoline through the body.
• Now, repeat "ah" several times, as if you are bouncing sound on the diaphragm. Find a speed that is comfortable to you.
• Imagine you are bouncing sound on each of the horizontal diaphragms: cranial, vocal, thoracic, pelvic floor, and the arches of your feet. Notice the changes in pitch.
• Move the sound anywhere in your body. Change the position of your face with each breath, allowing the sound to alter with the shape of your mouth.
• Add the body, inviting any movement that feels good, enjoying audible sound.
• Pause, in open attention, noticing the sensations of sound.

*Vocal warm-up exercise drawn from work by Claire Porter, choreographer, writer, and performer.

Sound score (alone or with a group)[*]

30 minutes

Seated comfortably, with paper and pencil, eyes open:

• Make a grid of eleven vertical lines across a paper, with each of the spaces between the lines representing one minute of time, going from left to right. Listen to sounds around you for ten minutes. Make up your own visual symbols for each of the sounds that your hear. As you hear them, place them on the page according to when they happened, how long they lasted, and how often they occurred. There will be layers of sound overlapping in a variety of ways. Some are ongoing, some are abrupt and happen only once, some repeat. You are creating your own notation so that you can remember the sounds and their qualities and re-create them later in your own way. 10 min.

• After the ten minutes, notice the rhythms of the events in your musical score. Using sounds you can make with your hands, mouth, breath, found objects, musical instruments, and so on, decide on a certain sound you will make for each symbol whenever it appears as you scan across the page from left to right. (Note: each sound you heard in the ten minutes will be represented by a different sound you'll be making. You may be able to play only a single layer or just a few layers at a time.) 10 min.

• Now play a rendition of it for yourself or someone else. 10 min.

• With a group, assign one person to each of the sounds, and re-create the whole pattern simultaneously. The time frame could be expanded or contracted, depending on how fast you read across your score. You might decide, for example, that it will take only one minute to play back the whole, or twenty minutes, instead of the original ten.

Place visit: Attention to sound and hearing

Experience your place through sound: Listen carefully to all the sounds at your place. Then choose one sound and listen attentively; give yourself time to invite associations. When you are ready, "listen in" on your internal response to sounds, and notice the process of interpretation and association.

Now consider all the aspects of your place that are listening to you, affected by your sound and presence. Notice how it feels to be heard. Remember: plants respond to human sound. Many animals, birds, and insects have more acute hearing than humans. Sound carries beyond the space you are in, linked by air, and your feet vibrate the earth. 20 min. Write about this experience. Read aloud to yourself or a small group, listening as you speak. 10 min.

[*]Exercise by composer/performer Mike Vargas.

DAY 13

Vision

Looking into the branches and trunks, between the things of the world, is a fundamental practice for opening our attentional focus, for stretching ourselves beyond our predominant view, our habits, our fearful projections.
—Laura Sewall, *Sight and Sensibility: The Ecopsychology of Perception*

The way we see shapes our view of the world. Seeing is a learned experience, developing and changing throughout our lifetimes. As with all modes of perception, our vision is affected by intention, which is influenced by personal goals and previous learning in the context of health and genetic tendencies. The process of seeing begins with the eyes, our primary sensory organ for gathering visual information from the environment. Visual perception also involves the brain, our primary organ for organizing and interpreting information from the body, including the other senses, our memories, and our associations. Thus, vision responds to both internal states and to the external environment; it is a translator between body and earth.

Vision is our dialogue with light. As light envelops and penetrates our bodies, the human eye responds to only a narrow band of wavelengths that can engage its photoreceptor cells without damaging tissues. This same middle-range spectrum is also used by other animals for information about the environment and by plants for photosynthesis.[1] In evolutionary history, most living forms have had light-sensitive structures. Early photosynthetic bacteria and subsequent plants made food from light. Watch the garden

Hubbard Squash, painting by Pamela H. See. Oil on panel.

sunflower turning its head to follow the sun and a windowsill geranium gradually orienting leaves and flowers toward the source of light. Through time, organisms evolved more complex structures for translating light waves into the electrical language of their nervous systems. Early insects, like dragonflies and cockroaches, evolved a compound eye that creates a rather grainy picture but is highly responsive to rapid movement. Notice their speedy reactions when you try to catch one! Vertebrates, including some mollusks and humans, have a "camera eye" that constructs a stable visual representation of the surrounding environment.

Over half of all the sensory receptors of the body are in the eyes. To accommodate this input, as much as 30 percent of the human cortex is devoted to processing visual information.[2] This dedication of resources reflects the evolutionary importance placed on vision for survival of our species. As our hominoid ancestors evolved in the dense jungle environment, brachiation and feeding required eye-hand coordination and favored acute, object-oriented vision over smell. Eyes gradually moved forward on the head for distinct focus, and muscles refined to focus two eyes on a single point, essential for hunting and gathering food. Many animals have even more acute eyes than humans have: bobcats can hunt rabbits in the dark, and ospreys can dive for salmon from hundreds of feet above a stream. Yet the complexity of the human brain, allowing interpretation and abstraction of visual information, is unique.

Color, that delicious component of everyday life, is limited among mammals to humans and some primates. White light contains all colors; the colors that we see are based on the wavelength of light—the longer the wavelength, the redder the light; the shorter the wavelength, the bluer the light. Our bodies respond viscerally to color; moods and health are influenced by the habitat around us. Foods are often selected or rejected for their color. Most of our mammalian relatives, including family pets such as dogs, horses, and cats, view the world in black, white, and varying shades of gray, depending on smell for refined discrimination. Color vision is a critical sense for most birds, many fish and reptiles, and insects, some of which utilize a different color range from that of humans. For example, hummingbirds and honeybees see ultraviolet; rattlesnakes see infrared.

The process of vision begins in the eye. About one inch in diameter, the eyeball is encased in the *sclera*, a tough but elastic white coat of connective tissue. The transparent, curved surface at the front of the sclera is called the *cornea*, the first and most important element in the light-focusing system of the eye. Light passes through the clear fibers of the cornea and is refracted through a watery fluid called the *aqueous humor*, which helps maintain the shape of the eye and nourishes the lens and cornea. As light enters the hole of the *pupil*, the doughnut-shaped colored membrane called the *iris* responds. Two sets of smooth muscles, circular and radial, regulate the opening and closing of the pupil, reacting to changing light levels: if light is intense, the iris contracts; if light is dim, the iris expands.

Light continues through the dense, clear fibers of the *lens*, which is convex in form. The lens has two sets of attached muscles to change its shape

PERSPECTIVES

Hiking in the Wind River Range in Wyoming in July, our family made a game of counting different types of wildflowers: inky purple star gentian, creamy white fringed parnassia, yellow heart-leafed arnica. Noting color, leaf, and petal patterns, we had come up with over one hundred different varieties on our three-hour walk when we encountered a lepidopterist staring into a field. He was looking for a rare species of butterfly. As he showed us the characteristics and we changed our point of reference, everything was different. Instead of looking for flowers, rooted to the ground, we were looking for movement, scanning the air. It was like dialing the lens on a kaleidoscope and seeing a whole new picture in exactly the same place.

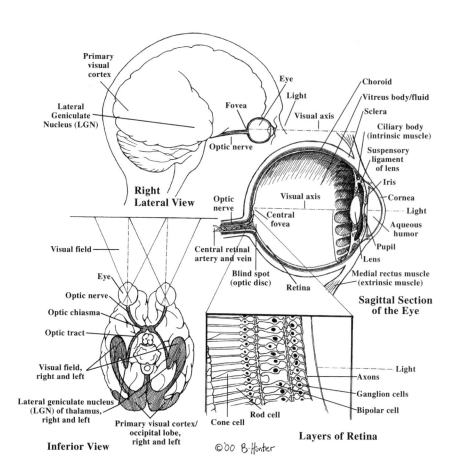

and light-focusing ability. The chamber behind the lens is filled with a clear, gelatinous material, the *vitreous humor,* comprising up to 80 percent of the eye's volume.[3] The *choroid,* a thin membrane lining most of the inner surface of the sclera, supplies blood to the rest of the eye. It also contains the pigment melanin, providing a light-absorbing layer that deters "extra" light from blurring the image.

The *retina* is where seeing begins, the beginning of the visual pathway. It can be seen by looking through the pupil with an ophthalmoscope and is considered to be the only part of the brain that is visible from outside the skull. Composed of hundreds of millions of cells, the retina is a thin tissue lining the inner surface of the choroid in the back three-quarters of the eye. (You can also directly view blood vessels in the interior landscape of the eye to screen for illnesses such as hypertension.) The cornea and the tightly packed fibers of the lens focus light rays onto a relatively small area of the retina, inverting the light so that the image projected is upside down and reversed left to right.

The *photoreceptors* in the retina are of two types, rods and cones. The 120 million *rod cells* in each eye are more abundant toward the periphery of the retina and are exceedingly sensitive to light. They allow us to see in dim light, such as moonlight, registering images in shades of gray. The 6 million *cone cells* function in bright light and are specialized for color vision. Individual cone cells contain one of three photopigments: yellow-orange, green, or blue, blending their information to produce all other colors. If all three are stimulated, we see white; if none are stimulated, we see black. Cone cells are especially abundant in the center of the retina, with two million packed into a tiny depression known as the *fovea*. Because of the high density of receptors in the fovea, we have detailed perception of color and form in the center of the visual field.[4]

Nerve connections from the rods and cones in the retina convert information into electrical impulses for the brain. A *blind spot* is created where these neural circuits converge to form the optic nerve, creating the only place on a healthy retina that can't respond to light. The *optic nerve* of each eye travels all the way to the *occipital lobes* at the back of the brain, the primary visual center, for preliminary processing. At this juncture with the occipital lobe, conscious recognition requires interaction with other "higher" regions of the brain, informed by memory, association, and expectation—our body history and worldview.[5]

Fluids bathe and fill our eyes. For protection of the outer, air-exposed surface, the eyes receive a constant rinse from the lacrimal glands (located in the socket) and other glands of the eyelids, which continuously secrete tears to keep the membrane moist and free from microorganisms. The reflexive action of opening and closing the eyelids pumps blood through the tissues, essential to the health of these complex sensory organs. The aqueous humor, which helps maintain the shape of the eye and nourishes the lens and cornea, must be constantly replenished; drinking adequate quantities of water nourishes the eyes in many ways. The eye socket itself, significantly deeper than the eyeball, is bathed in fluid, allowing mobility within the joint.

Six *extrinsic muscles* surround each eye, making movement possible in all directions within the bony socket—similar to the hip or shoulder joint. Binocular vision offers perception of volume and depth, a three-dimensional view of the world, by giving each eye its own unique view. It also requires entrainment, focusing two eyes individually toward the same image. Eyes move reflexively in a series of tiny stops and starts. During the pause between these oscillations, they process incoming visual data. Studies suggest that this movement occurs in frames of about a tenth of a second, which combines in our brains to form a seamless picture of the world. During these movements the eyes scan the visual field to pick up information and allow the small area of the fovea to gather details according to our intention at the moment.

Intrinsic muscles within the eyes adjust both the size of the pupils in response to changes in light and the shape of the lens in response to depth of focus. Both intrinsic and extrinsic eye muscles need to move to pump the oxygen-rich blood, remove waste materials, and balance useful tensile contraction. We can enhance or lose our resiliency at each level of this movement spectrum. Exercising the intrinsic muscles, which adjust the pupils in response to light; the extrinsic muscles, which scan the perceptual field for data and move the eyes in their sockets; and the muscles of neck and spine, which position the body for seeing, facilitates healthy vision.

We can direct our whole body with the muscles of the eyes. As with the

Before walking on Dog Island, near Appalachicola on the Gulf of Florida, I asked if there were snakes. "I don't think so," our hotel owner responded, and we set off to explore. We began along the gorgeous beach, then cut in through waist high marsh grasses in the wetlands, and finally came back along a dirt road past inland ponds. Mid-step, I saw a water moccasin sunning itself on the road about three feet away. We backed up quickly, then, from a distance, noted the diamond-shaped head and markings to be sure. When we got back to the main road, a resident said, "Oh, yes, they're all around, especially near the wet places." Expectation affects experience. This information changed where we walked and what we observed. Never again could we repeat that naive, relaxed stroll.

Once I was on a panel about perception with a visual artist who became blind late in life. His comments fascinated us all. One of the major losses he reported was a sense of scale. Touch gives you detail and is highly suggestive, but you don't know the context, the overall view. You can touch the corner of a sculpture and be totally inaccurate about the whole. Yet in the familiarity of seeing, it is easy to miss the details and associations evoked through touch. Take one sense away, and experience changes; every sense feeds the others.

"spotting" of a ballerina in multiple turns, the precise cast of a flyfisher, or the inspired catch of a baseball player, where we set our vision determines how and where we move. Changing focus is an increasingly necessary antidote to single-vision workplaces. If there is tension in the eyes or prolonged states in which they are used in a fixed position at a constant focal length—like looking at a computer screen or reading a book—the eye muscles and related muscles in the body become rigid. To exercise the eyes and the body as a whole we can focus up close on a leaf, far away at the horizon and trees, and in the middle range at other people passing by as we walk.

Visual perception is a linking together of multiple subsystems in the brain. As neurological signals begin their journey through the optic nerve, they pass through major relay stations, including the optic chiasma, where half of the axons for each eye cross to the opposite brain side, allowing each eye to see the right and left sides of an object and giving visual depth. Then they continue on through the optic tract to the lateral geniculate nucleus (LGN) of the thalamus, linking vision with all other incoming sensory information (where we merge with what we are seeing before the neural information even arrives at the visual cortex). Signals continuing on from the thalamus travel to the primary visual cortex at the back of the brain, our "mind's eye." There the image is literally mapped onto the surface of the brain's visual cortex, with an exact correlation between the locations of cells on the retina and those in the visual cortex. Stephen Kosslyn, a psychologist at Harvard University, calls this initial processing area the "visual buffer." Colors, shapes, line, shadows, edges, depth of field, and other aspects are first processed separately in distinct regions of the visual cortex to gather information, without actual visual recognition.[6]

Specific features of that information are then sent on to two higher brain centers for recognition, often referred to as the "what and where" systems. The "what" system (temporal lobes) compares the visual data with shapes and colors of other objects, and the "where" system (parietal lobes) considers spatial orientation and location. Next, the visual association centers (occipital lobe) may be engaged for visual memories relating present and past visual experiences and the frontal lobe association area for specific details and decision making. Through this process, we "recognize" the image in the visual cortex, binding all the sensory information into a unified whole. Information flows in both directions, with constant feed-back and feed-forward between lower and higher brain centers, so we are continually influenced by our own interpretations and decisions—we are responding to our own responses. If someone is blind and eyesight is returned later in life, the brain registers only patterns of light and dark, without visual recognition.[7] It must relearn, increment by increment, how to interpret the connection between the physical world and the patterns projected in eye and brain.

If we took in everything there is to see, we would be overwhelmed. Thus, within the visual system, there are several junctures at which there is a screening of neurological information to reduce the number of signals to be processed. The first screening occurs within the eye itself, in the retina. Although all visual detail from the fovea is sent on through the optic nerve to the brain, only one bit of data per thousand or so at the outer edges of the retina gets through perceptual gateways. A second screening for relevance occurs at the lateral geniculate nucleus. Selection favors pathways or signals that have been facilitated before—familiar visual stimuli. Habit can dominate our process of seeing. You see what you expect to see, find what you are looking for, and overlook what is new. The first time you visit a new place, for example, you will notice what you are most used to seeing, like the trees. Weeks later you may attend to the patterns of the bark, the unique insects that inhabit the crevices, or the nest tucked amid the branches.

Because seeing involves complex processing by the brain, it can be the slowest of the senses. When you look at a movie at twenty-four frames per second, it seems continuous. Yet vision is also the fastest sense, accessing quantities of information in a single glance for survival reflexes. We often fill in visual information, jumping to conclusions with partial visual cues. For example, we see a long dark object in the grass, and we think "snake"; we see a familiar red hat on a figure walking up a path, and we recognize "Mary." For speed of response, some visual cueing is essential for survival; other automatic reactions limit our experience. Part of visual training is noticing visual habits and allowing ourselves to take in more of what we see so that we can respond more accurately—interacting with, rather than reacting to, the world around us.

Visualization techniques utilize visual-neural pathways in reverse.[8] When an image begins in the cerebral cortex as a thought (imagine a tree), it travels down to the occipital lobe to the visual buffer. As we deepen our visualization (what does the tree look like, what is its smell, what movement does it inspire?) associations surface to fill in precise details, until the image becomes quite clear. In this way, we create a visual representation of our thoughts that can seem as real as an actual memory. When we visualize, imagine, or image an activity, we also stimulate neuromuscular circuitry, engaging a full body response, even when we are sitting still. (Imagine climbing the tree: where do you put your foot, where do you touch the branch?) If visualization is accompanied by actual physical practice, it is highly effective in stimulating efficient, integrated, neuromuscular circuitry. Armchair rehearsals are common to athletes and dancers trying to perfect their skills by visualizing the precise action they hope to accomplish. Visualization techniques are used in pain management, in relaxation clinics, and for increasing human potential in a variety of fields.

Each person perceives the world filtered through the matrix of his or her own life experience; no two humans see the same landscape. Through awareness of seeing and visualizing the world, we can begin to understand the many ways that vision influences our ideas and our actions, serving as translator between body and earth.

Snake, drawing by Laura Lee. Graphite.

TO DO

OF SPECIAL INTEREST: ECOLOGICAL PERCEPTION

Ecopsychologist Laura Sewall suggests five guidelines to enhance "The Skill of Ecological Perception," the title of her chapter in the text Ecospsychology: Restoring the Earth, Healing the Mind. *Look at something in the landscape around you and notice how your experience is affected as you (1) focus your attention (attend to the moment), (2) perceive relationships and context (notice connections between, looking both at details and at the whole), (3) maintain perceptual flexibility (relinquish expectations and judgments), (4) reperceive depth (include yourself as part of the landscape), and (5) invite creative vision (investigate your own creative imagination, imagery, and movement).*[9]

Changing focus
10 minutes

Standing or seated, eyes open, focusing on a specific object:
• Begin with an active gaze: project your vision out to the object to see it.
• Change to a receptive gaze: allow what you are looking at to come to you.
• Engage sharp focus: look at specific detail (engaging the fovea).
• Try a soft focus: allow your gaze to blur slightly so that you can see the whole rather than focusing on distinct parts.
• Open your visual field to include peripheral vision: look straight ahead but attend to information on the edges of your visual field.
• Scan distance: look at the horizon, something at middle distance, something close to you (engaging the intrinsic muscles of the eyes).
• See from the back of your skull, the visual cortex; include yourself in your field of vision.
• Notice which of the above patterns of seeing is familiar, which feels new.
• Now close your eyes for a few seconds, then open them to the light. Notice the iris and pupil adjusting. Dim light stimulates the light-sensitive rods in your eyes; brightness, the color-sensitive cones.
• Finish by moving your eyes; look up, down, side to side, on diagonals, and in a circle as you hold your head still (engaging the extrinsic muscles).
• Now hold your eyes at a fixed point and turn your head. (For example, turn your head left, as you look right.)

We often "fix" our eyes in their sockets, holding our eyes at a consistent focal length. When we do, the eye muscles get rigid, like tight shoulder or neck muscles. Movement brings blood flow, keeps the tissue elastic. Try these exercises a few times each day, stimulating responsive vision.

Snapshots (with a partner)*
30 minutes

Standing, dressed for indoors or outdoors, one person (photographer) with eyes open; one person (camera) eyes closed:
Note: Demonstrate this exercise with eyes open one time before you begin.
• The "sighted" person holds a partner firmly around the waist and by one arm, and practices walking safely (indoors or outdoors).
• The signal for the eyes-closed person to open his/her eyes for a "snapshot" will be a hand squeeze from the partner. As soon as the squeeze stops, the person will again close his/her eyes. In other words, the "exposure time" equals the length of the hand squeeze. After five snapshots, change roles and repeat.

*Title and exercise drawn from Alton Wasson of the Contemplative Dance Workshop.

To begin:

- The eyes-open person (photographer) guides the eyes-closed partner to something that catches the interest for a "snapshot."
- The photographer carefully positions the partner for the "perfect" shot, using touch to signal the partner to bend the knees, tilt the head, move closer to the ground or the object, and so on.
- As the photographer squeezes the partner's hand, the eyes-closed person (camera) opens his/her eyes, and closes them when the squeeze is released.
- Continue to another place. Take five snapshots.
- Change partners and do the same exercise.
- When a total of ten pictures has been taken (each partner has experienced five snapshots), return to the starting place. 15 min.
- Write about your experience and read aloud to your partner. 15 min.

Place visit: Attention to vision

Walking to your place, notice everything that is red. You are engaging the specific cone cells in your eye receptive to the color red. When you arrive, acknowledge everything red at your place. Then pause and witness your place, expanding your gaze. When you are ready, let yourself be witnessed by successive aspects of place, like a tree, rock, animal. 20 min. Write about your experience; read aloud to yourself or a small group. 10 min.

FARMSTORIES: REMEMBERING

I began to remember the stories about the farm when we bought the land in Vermont. I had traveled around the world, yet even when living facing the Sound in Seattle, gazing at the mountains in Utah, surrounded by the lush fields of Massachusetts, I had a tether back to the farm in Illinois, which gave me my sense of home.

It was when planting the two apple trees, one cherry, that I began to feel my roots grow down into the dense clay soil of Vermont. Every story has its ending, every intake of breath has its release, and there is always more.

Read aloud, or write and read your own story about vision.

II

Body and Earth

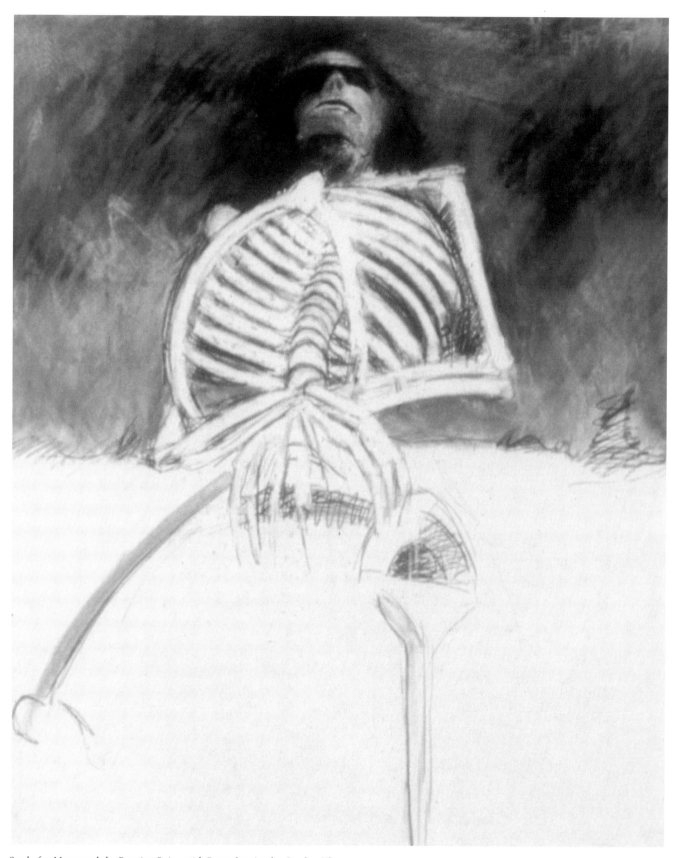

Study for Mona and the Burning Suit—with Boat, drawing by Gordon Thorne.
Oil and graphite on paper, 40 in. × 52 in.

DAY 14

Bones

Since we have left the water and become terrestrial creatures . . . we have added more and more solid features to support our containers of fluid upon the ground. . . . We had to develop our own geology, and learn to deposit our own bedrock of solid bone.
— Deane Juhan, *Job's Body: A Handbook for Bodywork*

Bones connect us to the ground. The skeletal system levers body weight toward the earth and gravity so that we can push away, allowing reach and stretch through air and space. Ease of bone articulation and efficient alignment help free muscular tension and joint range. Bones have multiple functions in the body. As well as transferring weight into the ground, they produce blood cells, store calcium and phosphorus necessary for the blood, support and protect organs, and provide attachment sites for muscles and connective tissue. Bones are dynamic, living tissue, changing moment by moment throughout our lifetimes. Bone shapes, including curves, hollows, shafts, and arches, allow both mobility and stability in movement and standing alignment. Because the body, like the earth itself, works as an interconnected whole, change in any one part affects the area above or below and can affect the entire body. When the bones are aligned, all other systems can function more efficiently.

Bones are our mineral body, similar in strength and mineral composition to marble.[1] The bones we see in classrooms and museums are the mineral salts that remain after death—mostly calcium and phosphorus—representing approximately 65 percent of bone by weight. The remaining 35 percent of living bone is a mixture of cells and connective tissue fibers in a fluid matrix, with numerous blood vessels and nerves, allowing rapid repair. In the human embryo, bones begin as cartilage and are gradually filled in during the first years. (Articular cartilage at bone ends and intervertebral disks remain in adults as evidence of this past.) Cells called osteoblasts continually build bone tissue, and osteoclasts continually break it down, balancing strength and lightness, depending on use and nutritional resources.

Compact bone forms the sturdy outside walls of long bones and the thinner surface walls of irregular bones. Inside, cancellous bone consists of a dynamic latticework called the trabeculae (formed of mineral salts), filled in by blood vessels and nutrient-rich yellow or red marrow. Increased production of trabeculae (such as in the shafts of long bones and the thick bodies of vertebrae) occurs in response to stresses of weight and muscular pull, reinforcing strength.

Bones are covered by a connective tissue sheath, called the periosteum

BASICS

My friend assembles skeletons of all sizes: tiny bones of mice, shrews, and voles detail her studio walls; a whale vertebra rests on the shelf. So when I find a moose vertebra in the woods, bleached white from summer sun, I think of sending it her way. But first, standing amid the balsam firs with fall scent in the air, I must take a moment to acknowledge this giant animal who shares my skeletal form and graces the land.

outside and the endosteum inside. This dynamic connective tissue nourishes bones and produces cells (which create osteoblasts) for bone growth. In function, periosteum is comparable to the nutrient-rich cambium layer of tree bark, supporting tree growth. Damage to the periosteum, like destruction of the cambium layer, can result in disease or death of bone or tree. In long bones, growth also occurs between the epiphysis (the end) and the diaphysis (the shaft), through cartilage along the epiphysial line. Bone growth and health are stimulated by compression and extension. Too much force causes bone damage; too little causes bones to atrophy or decalcify. It is important to remember that bones respond to use and abuse. Although humans share the same evolutionary form, each of us inhabits our skeleton in very different ways.

Connective tissue weaves the skeleton into an interconnected whole. Ligaments attach bone to bone, and tendons attach bone to muscle. Connective tissues are sufficiently interwoven that strong force on a tendon (like impact to the tough Achilles tendon at the heel) can result in broken bone. Articular cartilage covers the ends of movable bones, and joint capsules secrete and contain the lubricating synovial fluid between bones. The balance between the density of compact bone, the inner latticework of trabeculae, the interweaving of tough connective tissues, and the responsiveness of muscles and joints provides strength, lightness, and resiliency in our skeleton.

Bone and blood have a dynamic relationship. When calcium or phosphorus is needed in the blood, bone is dissolved; and when there is excess, they are returned to the bone for storage. Red blood cells (carrying oxygen to all the tissues), white blood cells of various types (serving as part of the defense system), and platelets (assisting in blood clotting) are all formed in the bone marrow and require regular replacement. Through the nutrient arteries and capillaries, blood constantly nourishes the organic tissues within bones so that every cell breathes. Oxygen and nutrients are absorbed and waste materials released through the semipermeable cell membranes.

From the cartilaginous notochords in primitive fish came the skeletal backbones common to our phylum, Chordata. The skull, spine, and rib cage form our central *axial skeleton,* integrating head to tail. The *appendicular skeleton,* which developed when amphibian ancestors moved to land, includes the peripheral shoulder girdle (clavicles, scapulae, and arms) and the pelvic girdle (pelvis and legs). By the time *Homo sapiens* walked across the African grasslands, the characteristic striding gait was initiated by a strike of the heel and push-off of the big toe for levering and propulsion, cushioned by an arch for shock absorption. A large skull housing the special senses and brain required rotation of the spine. A grasping hand, with thumbs long enough to reach the fingertips, increased articulation and manipulation, changing our relationship with the environment. Of the 206 bones in the human body, over half are in the hands and feet—approximately 26 bones in each, including wrists and ankles.[2]

Each joint type has a different function in the body. The shoulders and hips are ball-and-socket joints, the knees and elbows are hinge joints, the ankles and wrists form gliding and sliding joints, and fingers, toes, and jaw

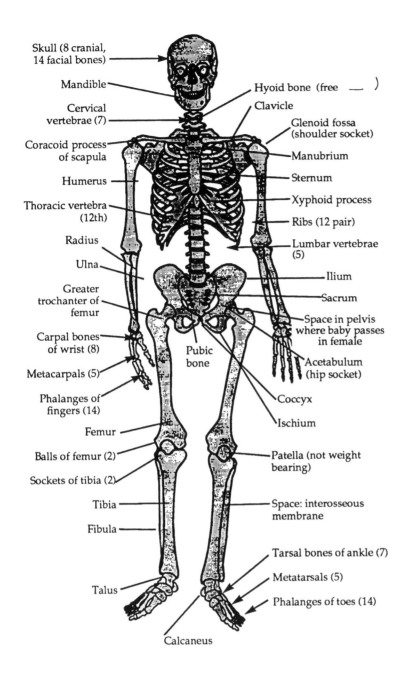

Skull (8 cranial, 14 facial bones)
Mandible
Cervical vertebrae (7)
Coracoid process of scapula
Humerus
Thoracic vertebra (12th)
Radius
Ulna
Greater trochanter of femur
Carpal bones of wrist (8)
Metacarpals (5)
Phalanges of fingers (14)
Femur
Balls of femur (2)
Sockets of tibia (2)
Tibia
Fibula
Talus
Calcaneus

Hyoid bone (free ⎯)
Clavicle
Glenoid fossa (shoulder socket)
Manubrium
Sternum
Xyphoid process
Ribs (12 pair)
Lumbar vertebrae (5)
Ilium
Sacrum
Space in pelvis where baby passes in female
Acetabulum (hip socket)
Coccyx
Ischium
Patella (not weight bearing)
Space: interosseous membrane
Tarsal bones of ankle (7)
Metatarsals (5)
Phalanges of toes (14)

Pubic bone

Anatomical illustration from LifeART.

are composed of small hinge joints. The spine positions the structure for movement, the ball-and-socket joints determine direction of movement, the hinges of elbows and knees extend range, the sliding and gliding joints of wrists and ankles provide shock absorption, and the small hinges of fingers, toes, and jaw allow articulation and manipulation.[3] Other joints within the structure add dimensionality to movement, like the pivot joint at the elbow, which allows rotation of the forearm, turning the palm up or down for increased adaptability.

* * *

The eight bones of the *skull,* are connected by slightly movable *sutures,* protecting the brain lobes during impact. The *frontal bone* covers the forehead, two *parietal bones* form the right and left top sides of the skull, two *temporal bones* surround the ears, and the *occipital bone* forms the rounded back of the skull. Inside the skull, a butterfly-shaped *sphenoid bone* traverses the center, forming the base for the brain and the ceiling for the mouth. (Open your mouth and touch the inside top surface—the wings of the sphenoid bone.) The *ethmoid bone,* forming the nasal septum, is one of seven bones forming the sockets for the eyes. (It can be touched by pressing your thumbs lightly into the nose side of each eye socket.)

Bony landmarks of the skull include the *occipital condyles,* the two "feet" of the skull, where the thirteen to twenty pounds of the head passes into the first vertebra of the spine. (This joint is directly in from the ear holes and can be located by nodding your head "yes.") The *occipital foramen* is a hole in the base of the occipital bone, through which the spinal cord passes. (Weight passes through the condyles *in front of* this hole, so there is no compression on the brain stem.) A bony recess in the center of the sphenoid bone, the *sella turcica,* supports the pituitary gland, suspended from the hypothalamus of the brain.

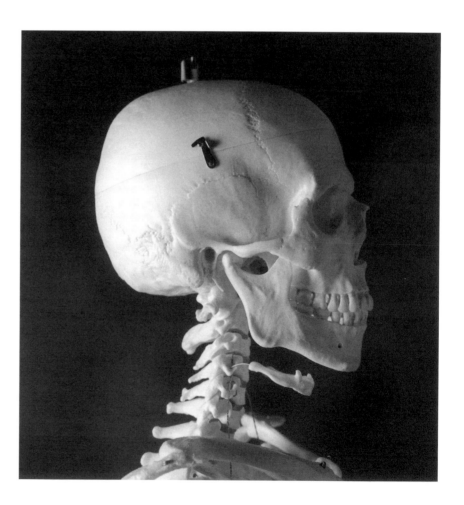

Skull, temporomandibular joint, hyoid bone, and cervical vertebrae, right lateral view. Photograph by Erik Borg.

Sphenoid bone with sella turcica. Photograph by Erik Borg.

The face is composed of fourteen *facial bones,* underlying our specific features. Of these, the *maxilla* forms the upper mouth, the *mandible* forms the jawbone, and the two *zygomatic bones* form our cheekbones. The highly mobile *temporomandibular* joint (TMJ) between the temporal bone of the skull and the mandible of the jaw, is a fluid-filled synovial joint with a cushioning disk. (Open and close the jaw; the masseter muscle, which moves the jaw, is proportionately the strongest muscle in the body.)

The *spinal column* is a series of reversed curves, beginning with the top vertebra, the *atlas,* and continuing to the tailbones, the *coccyx.* Intervertebral disks form the core of the undulating column. Each bony *vertebra* is composed of a solid *body* on the front (spatially replacing the cartilaginous notocord) for weight transfer with their interwoven disks, plus a *spinous process* and two *transverse processes* (which grew around the spinal cord for protection) on the back, forming the vertebral arch—our backbone. (We can see and feel the vertebral arch as the "bumps" of our spine.) There are seven *cervical vertebrae* in the neck, twelve *thoracic vertebrae* in the thorax (rib cage), five *lumbar vertebrae* in the lower spine, five fused vertebrae forming the *sacrum* (linking the two pelvic halves), and two to four individual or partly fused vertebrae in the *coccyx,* forming the forward curve of our ancient tail. The delicate *hyoid bone,* suspended by ligaments from the front of the neck, is an important base for the tongue. The spinal column works collectively, with movement equally distributed throughout the spine.

The *thorax,* or rib cage, is composed of the *sternum,* twelve pairs of movable *ribs,* and the twelve *thoracic vertebrae,* protecting the vital heart and lungs. The *sternum* is the broad flat bone at the front of the chest—it has three components, the manubrium, sternum body, and xyphoid process— with slightly movable joints between each to allow responsiveness in breathing. The top seven pairs of "true ribs" form movable joints directly with the

Our Brittany spaniel scratches at the door to get out, his mind ahead of his paw's limitations. Grasping the knob, I add a quick twist of my wrist, and he bounds through. A neat pile of white paint on the floor marks his efforts. In this moment, I recognize the poignancy of our human hand, appreciating one remarkable aspect of my form. The opposable thumb allowed our hominoid relatives the capacity to climb, feed, groom, and communicate with gestures. Tools were an extension of the hand, facilitating evolution of the brain and language. Grinders, arrowheads, and even doorknobs require planning, memory and social interaction—a heritage of thought still expressed in my fingers.

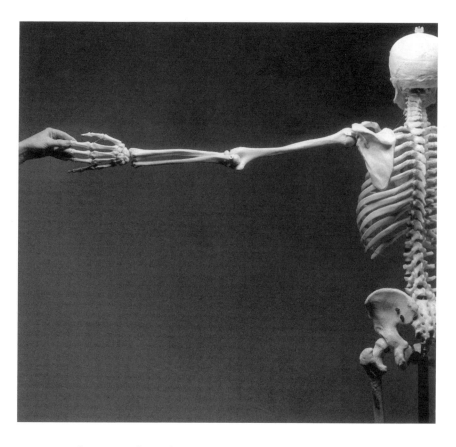

Shoulder girdle, ribs, spine, and pelvis.
Photograph by Erik Borg.

sternum; rib pairs 8 through 10 connect to the sternum through cartilage, and rib pairs 11 and 12 are "floating ribs" that end in the muscular sheath of the abdominal wall. Each rib pair connects to the body and transverse processes of one thoracic vertebrae, ribs 2–9 also lever into the body of the vertebra above, ensuring collective movement of the spine. In breathing, the ribs swing laterally and upward, making more room for the lungs to fill with air. Together, these three body areas, the skull, spine, and rib cage, form the axial skeleton, our ancient fish body.

The *shoulder girdle,* part of the appendicular skeleton, is generally free from weight bearing in standing postural alignment. In evolutionary development, these bones wrapped around the ribs, to give leverage and directional mobility as our ape ancestors swung through the trees. The only bony connection between the shoulder girdle and the older axial (fish) skeleton is the mobile *sternoclavicular joint* between the *manubrium* (breastbone) and the two *clavicles* (collar bones). The clavicles on the front surface, and the two triangle-shaped *scapulae* (wing bones), curving from front to the back, lie on the upper ribs. The *humerus,* the upper arm bone, forms a joint with the scapula above and with the *radius* and *ulna* of the forearm, below, transitioning force to the periphery. The radius (the long bone on the thumb side of the lower arm) can cross over the ulna (on the little-finger side), increasing mobility. Eight small *carpal bones* comprise the wrist, five *metacarpal* bones span the palm, and fourteen *phalanges* form finger bones. The mobile thumb has two phalanx bones, all the other fingers have three.

Each half of the *pelvis* is composed of three bones, which fuse in the human in the first years of development. The *ilium* fans out from the sacrum (the triangle-shaped base of the spine) on each side, forming the iliac crest that we feel at the top of our hips. The *ischium* descends in a series of curves and notches toward the earth on each side, forming the two "feet" of our pelvis—the ischial tuberosities, or sit bones. The *pubic bones* reach diagonally forward, like the prow of a ship, forming the front rim of the pelvic bowl. In each pelvic half, all three bones equally form the hip socket, the *acetabulum*, where the leg comes into the pelvis.

There are six movable joints in the pelvis, providing shock absorption during walking: two *sacroiliac joints* (connecting each ilium to the sacrum), two *acetabulums* (connecting each pelvis half to a leg), and two *interpubic joints*, also called the pubic symphysis (connecting each pubic bone to the connecting disk). Thus, the pelvic girdle and sacrum create a bowl, with a diamond-shaped opening formed by the four bones in the base of the pelvis. (Rock forward and backward on your chair to feel this diamond-shaped support.) In females, this "hole" is a passageway for the baby in childbirth.

The *femur*, or upper leg bone, extends horizontally away from the hip socket (you can feel the protrusion of the femur in your hip below the iliac crest), then diagonally down the *shaft* to the knee. The knee is formed by the two *balls* that shape the end of the femur connecting with the two *sockets* that shape the end of the tibia, the shinbone. Two disks, or *menisci*, cushion the joint within its fluid-filled capsule, thick at the outer edges and thin in the center to deepen the sockets. Four ligaments—two lateral (one on each side) and two cruciate (which cross in the center of the joint)—protect against impact and twisting. The *patella*, a small bone inserted inside the patellar tendon in front of each knee, serves to protect the joint from impact and creates a powerful pulley for the quadriceps muscle. In efficient alignment, the weight of the body transfers directly through the center of the joint, not the patella. (Stand and feel your own knees, balancing the bone ends within the joint.)

The *tibia*, the long lower leg bone, often the straightest bone in the body, transfers body weight to the inner ankle; the *fibula*, the thinner outer leg bone, braces laterally below the knee and continues along the outside of the lower leg to the outer ankle. There's space between the tibia and fibula, woven together by a tough connective tissue called *interosseous membrane* (like webbing in a tennis racket) and giving support and lightness to the lower extremities. Both bones rotate slightly with each step, increasing blood flow to the tissues. The bottom of the tibia and fibula create the top of the ankle joint, called the *inner* and *outer malleoli*. They are landmarks in postural alignment, as well as forming the upper arc of the ankle. Two-thirds of the joint surface at the ankle is tibia; one-third is fibula, articulating with the *talus* below, the first of seven tarsal bones of the ankle.

The *talus* moves forward and back in the "cave" of the malleoli, creating a hinge joint to flex and extend the ankle. The *calcaneus*, or heel bone, the largest tarsal bone, is shaped like a doorknob, giving leverage to the back of the foot. In standing postural alignment, weight passes equally through the

Visiting the American Museum of Natural History in New York, I stood on the top floor next to Tyrannosaurus rex, the "tyrant reptile," admiring forty feet of bone. This magnificent assemblage represented sixty-five million years in time and reminded me of the summer my older stepson kept dermestid beetles. They neatly devoured flesh off bones of the dead animals he'd find near our home: squirrels, a beaver, once a snapping turtle. After they'd cleaned off his specimens, he would feed the beetles bologna. When he broke his leg, feeding them became my job. I appreciated their attention to bone, I just hoped they never got loose!

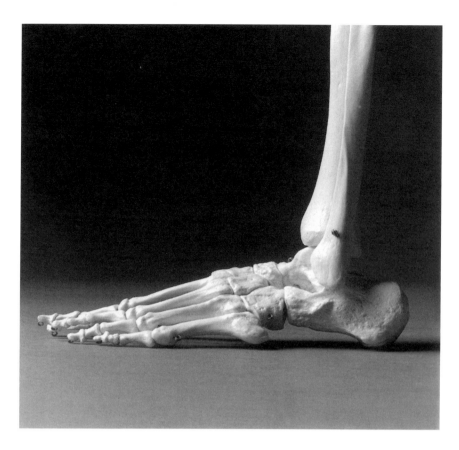

Left foot, lateral view. Photograph by Erik Borg.

seven *tarsals,* five *metatarsals,* and fourteen *phalanges.* Imagine an *X* connecting the big toe to the outer surface of the heel and connecting the inner edge of the heel to the little toe, for each foot. Stability and mobility of the feet come from equal distribution of weight through all the bones. If we tense our muscles, we imagine that we rely on ourselves for support; as we release our weight to the ground, we are supported by the earth![3]

⚛   ⚛   ⚛

TO DO

Axial skeleton (alone or with a partner)

10 minutes

Seated upright in a chair or standing:
• Begin the "yes" nod with your skull to find the connection between the occipital condyles and the first vertebra.
• Let the skull fall forward to initiate the curve of your neck.
• Continue rolling down the seven cervical vertebrae of your neck, one by one until you reach the base, the top of your ribs.
• Pause and take a deep breath. Continue rolling down the twelve thoracic vertebrae, connected to each rib. At the base of your rib cage, pause to breathe.
• Continue rolling down your five lumbar vertebrae, your lower back, connecting ribs to pelvis. Pause and breathe into this area. Be sure your head is hanging loose.

- Then release your sacrum and coccyx forward.
- Pause, hanging your head and spine toward the floor.
- Move your pelvis side to side and feel your head swing in response.
- Pause and reverse the whole process. Connect your weight into the chair or floor, and return to vertical alignment, bone by bone. Continue your awareness up to the sky and down to the earth, through extended proprioception.
- Repeat several times at different speeds (it gets easier), feeling the counterbalance between pelvis and head.

    With a partner: touch the back of each vertebra (the spinous process) one by one as you roll down and roll up, to identify each bone.

Moving all the bones

15 minutes

Lying on the floor, eyes closed (add music if you like):
- Begin with your feet. Explore any movement that feels good to your toes and feet at this moment. Follow images, ideas, emotions, or sensations in this area of your body.
- Allow the movement to travel to your knees. Explore how the bones of your knees connect and move, noticing impulses for movement.
- Notice how the movement of your feet and knees connects into your pelvis. Explore movements of your hip sockets. You may want to change your relationship to gravity: kneel or stand, feeling the joints take weight.
- Follow the movement of your pelvis up into your spine. Play the scales of your spine, exploring movement, large or small. Remember the three-dimensionality of your spine, allowing rotation, flexion and extension.
- Include the ribs; how do the ribs want to move? Each of the twenty-four ribs (twelve pairs) has several movable joints; explore possibilities of the rib cage.
- Extend the movement into the shoulder girdle: the scapula, arm, wrist, hand, and fingers. Explore movement impulses in these body parts.
- Bring your awareness to your head, as it connects to your spine. Remember, the skull can take weight, create momentum, and move in any direction.
- Add vision, and allow the joints of your eyes to participate in movement. As your eyes lead, your body can follow.
- Now pause in open attention. Return your full body to the floor and begin to rock the heels of your feet, jiggling the skeleton. Include all the parts: let your skull rock on the atlas, your ribs articulate on the spine and sternum, your legs move in their pelvic joints. Relax your jaw (TMJ) in the temporal bones and your eyes in their sockets; let them rock with the rest of your body.
- Pause in open attention.

Authentic Movement project. Photograph by Erik Borg.

Place visit: Attention to bone

As you walk to your place, notice the transfer of weight through your body to the ground. Focus particular attention on the push of the toes, the reach and strike of the heel. At your place, take off your shoes and massage your feet, stimulating the bones. Begin a walking meditation (shoes on or off). Moving very slowly, be aware of each step; imagine weight pouring through the bones to the earth. Experience place from this dimension of your body. Remember, bone is living tissue, changing and responding with every breath. 20 min. Write about your experience. 10 min.

CROSSING OVER (1996)

Change is a process, each stage of life its own season. "Do you recognize yourself at this age?" a friend asked. And I didn't. But I began to notice the details. Hands learn patterns and gestures throughout a lifetime, the skills of a trade. My father can no longer open any package, fit a key into its lock. His fingers trace the edges of door frames, making a determined effort at departure. "I just don't want to die fast," he says. And he took seven years.

"How do we get from here to there?" he whispers in my ear. "I want to go there and there and there." He is pointing at the pond, path, and road beyond the fence, beyond the locked gate where we stand. "Sometimes the door is open; sometimes it's closed," I whisper back. "Right now it's closed." And so he waits, until he can find his own way across quiet waters.

In my mind, a boat floats away from the shore. I am sitting on a log, listening to the waves. He raises his hand as a sign, and I know. Now that he has gone, I make dances, plant daffodils in the garden, stand a marble slab on its end. The land already holds his footprint as memory, marking a boundary between what was and what is to be.

Each of us has passages to make, thresholds to cross, our own rituals of change. Cycles interweave, connecting bone to soil, breath to air, tears to the glass of water we hold in our hand.

Read aloud, or write and read your own story about bones.

DAY 15

Soil

I have seen so many delicate shapes, forms, and colors in soil profiles that, to me, soils are beautiful.

—Hans Jenny, "A Life with the Soil," *Orion Magazine*

Soil is the skin of the earth. This living, breathing surface layer provides nourishment and support, as well as a final resting place for all living creatures—plants and animals. There is more living biomass below ground than above it, soil specialist Hans Jenny reminds us: "The soil contains over a thousand different species of lower animals, the earthworms, pill bugs, nematodes, millipedes, termites, ants, springtails, and amoebas, not to mention the millions of molds and bacteria. Soil organisms consume oxygen from the soil air, give off carbon dioxide, creating a multitude of respirations which characterize the metabolism of a soil individual." Considering soil as a living ecosystem helps us to understand its vital effects on our lives.[1]

Attitudes about soil vary, but most of us tend to ignore the crust of the earth unless called into direct relationship as farmers, gardeners, builders, or educators. Yet we are made of the same minerals that comprise the soil: 65

The Civil War Battlefield of Petersburg, Virginia. Collage (photograph and materials found at the site) by John Huddleston.

*Making dirt is a favorite occupation of gar-
deners, testing it between fingertips like pie
dough, looking for the perfect consistency.
Not too dry, not too dense, a healthy mix
of air, minerals, and organic matter. On our
garden of glacial till in Maine, we add
llama manure from our friends' farm. With
this fecund mix, the garden explodes into a
collage of sunflowers and pumpkins, spin-
ach, carrots, and herbs. On dry days we
water the lettuce, watching drops disappear
into the well-mulched earth. We use straw
as cover, to be tilled in next spring, feeding
the soil that feeds us.*

When I first performed Farmstories, *the editor of* The New England Review *asked if
the magazine could publish the text. I was
delighted and sent the manuscript. When
looking over the galley proofs, I realized
that the story on "Soil" was missing. "Oh,"
the assistant responded, "we wanted all the
stories except the one on dirt." A student
who was teaching fourth-graders about soil
helped me out with this distinction: "Soil is
what you find in your garden; dirt is what
comes in the house on your shoes."*

percent of our bone is crystallized mineral, with a content comparable to
many sedimentary rock formations—principally calcium and phosphorus,
with traces of magnesium, sodium, carbonate, citrate, and fluoride.[2] The remaining 35 percent is organic tissue nourished by nutrient molecules that
originate in the soil. Perhaps it is our ambivalence about the cyclic processes of digestion and elimination or about death returning components of
our bodies to the soil that makes us avoid, denigrate, and potentially destroy this essential ground. Yet the processes of decay are essential to recy-
cle energy and nutrients to support new growth, perpetuating life.

Although myths throughout time have related our origins to the dirt or
clay of the earth (the name Adam, in Hebrew, means both dirt and man), the
very word today has negative connotations: to be soiled, to be dirty. Even our
feet, which connect us moment by moment to the ground, are often considered disgusting. When we recognize that these vital and sensitive appendages
that support our weight contain a map of the whole body through reflex
points for all the vital organs, perhaps we can recognize that the soil, too,
connects us to the larger web of life on earth and supports our existence.

Approximately 99 percent of the world's soils develop from the weathering of bedrock, the solid rock of tectonic plates beneath loose soil. The remaining 1 percent of soil is developed through decay of organic materials
such as plant and animal matter. Bedrock weathering produces a basic soil
material that evolves further into different layers, called horizons. The first
horizon to develop is the organic topsoil. As soon as there is enough soil,
plant species, beginning with tiny lichens, colonize the area. The growth of
plants attracts animals, bacteria, and fungi. When these organisms and
plants die, their organic matter is incorporated into the soil, to form humus.

Humus is the organic matter in soil, the rich dark material essential to
soil fertility formed by the waste materials and decay of plants, animals,
bacteria, and fungi. Humus can be present in any proportion; it contributes
to the scent of particular soil types and is vital to soil's life-sustaining abilities. It holds the mineral particles together and gives soil its spongy texture,
which allows it to retain the water and nutrients needed for plant growth.
Humus also maintains space for fragile root hairs, encourages extensive
root systems that stabilize the soil, and promotes gas exchange between
below-ground roots and the atmosphere.

The parent material continues to weather from below, and humus continues to accumulate from above. Eventually, one or two new layers form
between the topsoil and the parent material. Percolating water leaches minerals and small particles from the surface and deposits them in a new horizon.
In forest soils, there may be a mid-horizon, where silt and sand accumulate in
a light gray layer. This light gray area indicates that nutrients pass quickly
from the upper to the lower soil layers, until they reach bedrock, the soil
foundation. You can carefully dig a soil pit and observe the various soil horizons, returning the soil and restoring plants when observation is complete.

When bedrock weathers, clay minerals and quartz usually play important roles in the dynamics of plant growth. Clay minerals increase absorption of water molecules and plant nutrients, such as potassium and cal-

cium, in the soil by forming ionic bonds with these elements. Quartz crystals function as catalysts for keeping the soil loose and aerated by forming sand grains, which allow good drainage. Farmers and gardeners generally prefer topsoil containing rich humus mixed with a composite material consisting of equal amounts of sand, silt, and clay for nutrients and drainage.

Soil is the foundation of the food chain. The tiny soil organisms responsible for the storage and recycling of basic elements crucial to the existence of all living things are microbes. Present in the land and the sediments of the ocean, microbes decompose dead plant and animal tissues as well as absorbing carbon, nitrogen, and phosphorus from the ground and air. When they die, microbes release these elements back into the soil or the ocean floor. Some carbon and nitrogen diffuses into the atmosphere, while the rest is left for uptake by other microbes and higher organisms such as plants. Nutrient transfer from microbe to plant begins the movement of the earth's elements from cell to higher cell, including the cells of our bodies.[2]

Eventually, we ingest soil nutrients through the foods we eat, such as the grains and fruits in our breakfast cereals, participating in the continuity of larger cycles. Plants fuel all other species and, in the process of decay, nourish the soil itself. Earthworms, ants, and small burrowing animals such as moles and chipmunks, help with soil aeration and increase fertility. Plant roots such as those from grasses and trees stabilize the soil, preventing erosion of fertile topsoil by wind and water. Humans are responsible for the health of the earth in the places we inhabit: what we put in or on the soil ends up in us. Caring for soil is an ongoing process affecting water and air, flora and fauna, and the health of the living, breathing ecosystems we call home.

Soil itself is delightfully sensual. Variations in color, smell, shape, texture, and layers evoke depth rather than surface perspective for those who take time to observe. Like skin, soil has layers, some more nutrient-rich than others. Soil characteristics vary with geographical location. The size of mineral-containing particles helps determine the soil profile and its name. From smallest particle to largest, soils may be called clay, loam, silt, sand, or gravel. The tiny clay particles that bind tightly to each other we call "heavy soil"; "light soil" is composed more loosely, of sand and silt. A soil profile reflects the integrity of a soil ecosystem as well as the proportion of its components, from bedrock to surface layers. Most soils are about 50 percent minerals, plus varying combinations of air, water, and organic material. In each soil profile, many species make the soil their home.

Soil erosion, the removal and loss of soil by the action of water, ice, gravity, or wind, occurs normally in all systems, but at a rate which is counterbalanced by the emergence of new soil from the subsoil. Aldo Leopold, in *A Sand County Almanac*, provides a useful perspective on erosion: for every atom that is washed from the prairies into the sea, another atom must be pulled from the bedrock. In areas where vegetation is removed (resulting in a loss of protective covering, humus, and root networks to hold the soil in place), particles of soil are more susceptible to the flow of water or wind. Overgrazing, poor crop practices, clear cutting or massive burning of forests

On our Midwest farm in the fifties, it was my job to dust. Each Saturday, I would wipe a layer of soil off all exposed surfaces. The rich loam that gave us our livelihood was blowing in the wind. Erosion practices in those days were being sacrificed for yield: hedgerows and windbreaks cut for more footage, cover crops omitted to get multiple plantings. Now hedges are being replenished, new planting methods developed. When our once vast prairie was plowed in the mid-1850s, they used a device developed in New England by John Deere in my hometown, Middlebury, Vermont. Wooden, then metal blades broke open the soil, exposing it to wind and water. As European settlers moved west from the newly depleted soils of New England, they replaced the nomadic lifestyle of previous caretakers of the land, small bands of Illini, Iowa, and Cahokia Indians. Today both of these landscapes are in recovery. Maybe stewardship of place will help keep our rich soil in the fields, instead of in our houses.

When we moved to Vermont, there was a remnant pond behind our 1830s farmhouse. One day a neighbor said, "Your pond was the local swimming hole in summer," and Sonny came with his backhoe to patiently redistribute the earth. As it drizzled rain, the Vermont clay banks slid into the slowly widening hole; displaced soil oozed like molten lava toward the water willow upstream. The next day, this is where we found our sons and their friends, immersed waist-deep in clay. "Spa Middlebury," we called out, enjoying the sight, and they eventually flung their soil-coated bodies into the now-full waters. I too slipped down the bank and submerged, slathering arms, chest, and face to complete the effect. Amid earthen smells and primal sensations, I remember cool mud on hot, relieved skin, ten-thousand-year-old lake bottom in my hair.

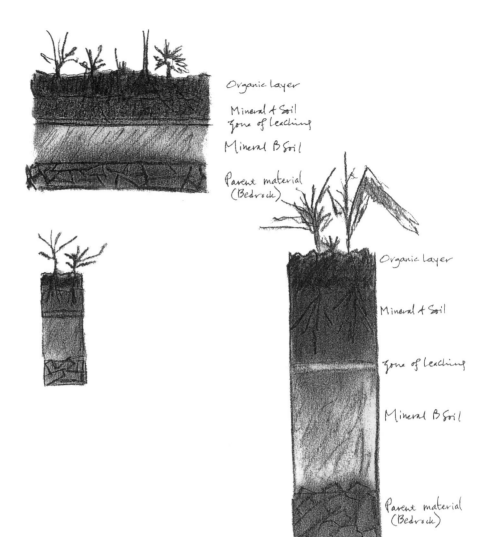

Soil profile. Drawing by Eve-Lyn Hinckley.

"Dinosaur Footprints," the sign announces, and I follow the winding path down the steep shore of the Connecticut River. There, embedded in slabs of sedimentary rock, are telltale prints of three-toed dinosaurs, Eubrontes, who walked along the mud flats of a lake some 190 million years ago. Perhaps the prints were covered by sediment-laden floodwaters, with repeated deposits solidifying and preserving the impressions. You can still see the toenails pressed into damp soil, imagine the blood pulsing in their veins.

are major causes of soil erosion on our planet today; desertification from depleted soils is a growing threat worldwide.

Erosion often couples with contamination: materials that have eroded from the soil pollute water reservoirs, and silt builds up in rivers, causing flooding, increased erosion, and loss of important habitats such as stream banks. Pesticides, herbicides, and fertilizers kill important insect and animal life that contribute to the health of soil and also pollute ground water. Generally, if you lift a rock or dig a hole and no earthworms or beetles are present, the soil has been poisoned. Reducing livestock herd sizes, and restoring our forests and farmlands through sustainable practices are necessary for soil health.

"We need a kind of reverence for the soil, for the creature world inside it, and for its character expressed in the profile features," Hans Jenny encourages. "Where big logging equipment turns soil upside down to make earth beds for falling redwood trees, the mass of soil remains at the site and no 'environmental damage' is said to occur. Yet the soil profile, the soil's signature and identity, is obliterated. Though I consider such profile destruction an irretrievable loss, I have never seen anybody shedding tears about it."

Consider all the things in your house derived from the earth's crust: there may be rock walls unique to your region or stones gathered from a trip. Soil nourishes roots of the potted plant, or travels on the bottom of your shoes. There are nails, bolts, and metal of all sorts; concrete; and wood, fabrics, and paper from trees that grew in soil. You may have jewelry with gemstones, silver and gold fittings; glass windows are melted sand. There are minerals in your bones, from the Midwest soil perhaps, transported through the grains in the bread you ate for breakfast. You can imagine the soil beneath the building, the bedrock under the soil, the tectonic plate of this particular continent floating on a molten layer of rock, and the solid, iron core of the planet.

Whether we are living in cities, farms, or the wilderness, soil is fundamental to human existence. "Human bodies belong to and depend on dirt. We spend our lives hurrying away from the real, as though it were deadly to us. But the soil is all of the earth that is really ours," William Bryan Logan reminds us in *Dirt: The Ecstatic Skin of the Earth*. It is important to understand how nature creates soils and how long it takes to make a soil. We can locate soils of known ages that extend over decades, centuries, and millennia, providing the ground we walk on, the food for plants upon which we depend for food and oxygen, and the beauty and diversity that nourishes our senses. Ultimately, our bodies return to the soil. Loss of soil can be viewed as loss of self. In Hans Jenny's view, "Healthy soils make healthy people."

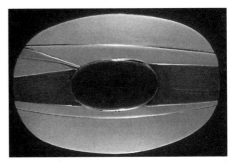

Belt buckle, by Harriet Brickman/Tom Raredon, from *River Series*. Sterling, 14-karat gold, lapis lazula, 2.25 in. × 3.5 in.

Reading the signs: ten questions to ask the soil
2–4 hours

TO DO

Consider the soil where you live; begin by asking these questions:

1. What are you made of? (Notice protruding bedrock, the parent material of soil, as well as the proportion of pebbles, sand, silt, and clay; notice the vegetation, animals, and microscopic organisms forming the decaying organic material of the humus.)

2. Do you have specific geological markings here or nearby? (Consider mountains, road cuts, faults, cliffs, and waterfalls.)

3. Do you show signs of glacial activity? (Look for boulders in the middle of fields or woods, hills comprised mostly of sand and gravel, glacial ponds or potholes, scoured rocks.)

4. What grows in you, here and nearby, naturally and commercially? (Notice trees, grasses, crops, orchards; watch for telltale insects and birds.)

5. Do you show signs of erosion by wind or water? (Consider muddy waters, rounded mountains, barren hilltops, washed-out road gullies, and dust in the air.)

6. What rocks do you contain? (Identify two, using a field guide; note if

they are igneous [formed from fire], sedimentary [formed under water], or metamorphic [formed under pressure].)

7. What do you do about water? (Consider slope, rivers, wetlands, and ponds; note the textures [dry or saturated]; note soil implications for drinking water; consider drainage ditches, culverts.)

8. What animals call you home? (Notice domestic and wild mammals, birds, amphibians and reptiles, insects, and humans. Include those you can see and those recorded by others.)

9. What's your relationship to humans? (Reflect on forestation practices, agriculture, industry, architecture, landscaping, roadways, and litter. Consider indigenous peoples and changing populations; artifacts, cellar holes, fences, dumps, and burial grounds.)

10. What is your aesthetic profile? (Consider color, smell, contour, rhythm, and texture.)

Savasana: Corpse pose
5 minutes

Lying on the floor or ground, arms slightly away from your body (palms up), legs comfortably apart, eyes closed:
• Allow the weight of your head and body to spread into the ground.
• Imagine the skin softening and melting away from your face and jaw, your torso and arms, your pelvis, legs and feet, softening toward the earth.
• Imagine the muscles releasing their grips on your bones, dropping away and melting into the ground.
• Imagine the bones of your body relaxing into the ground, connecting to the earth. Let the bones of your face and neck, ribs and arms, torso, legs, and feet gradually dissolve back into the ground.
• Notice your breath, full, connecting with the air around you.
• Allow yourself to rest in this state, breathing deeply, noticing sensations.
• Gradually roll to one side and transition to seated.

Savasana is a yoga pose for deep relaxation. There are various ways to guide the journey, allowing the body to be supported by the earth, then dissolving distinctions between body and earth.

Place visit: Attention to soil

As you walk to your place, focus awareness on the rhythm of the topography of the land. Notice how it curves and changes, how the vegetation is affected, and how your body responds.

At your place: Bring your attention to the soil. Invite full participation of the senses. Dig, touch, and look carefully, using the ten questions above to guide your reflections. Imagine the living, breathing activities of soil as the skin of the earth. 20 min. Write about your experience. 10 min.

Yoga: Corpse Pose, Savasana.

CROSSING OVER: HOLES

I crawl on hands and knees in the backyard, digging bulbs. My parents are moving; I may not return to this place. My fingers clutch at dark soil, gathering fat, pink-tipped shoots of rhubrum lily, my birth flower, to transport to my home. I want to take it all: big mounds of earth, tubers, worms. No suitcase will hold all that I desire to carry with me. I give up, turning to gather leaves into a pile by the curbside.

My father wanders to the backyard and begins raking. "What do you have there?" he asks, worried about the holes I have dug. "I don't want to fall in." Covering each one, he tramples soft soil with unsteady feet until firm. Lifting the empty bowl off the birdbath, he places it over the exposed earth.

There are holes all round: holes in drawers where cameras and photographs have been taken away; holes in the basement where his mother's diaries were once hidden; holes by the workbench where the shovels and pitchforks have been sold; holes in the mind, where one thought connects to the next; holes in the space between us, where I reach out, looking for a familiar hand.

Read aloud, or write and read your own story about soil.

DAY 16

Breath and Voice

The pharynx is our instrument of expression—of communication and relationship between ourselves and others, from our deepest knowing and intuitive self to knowledge which lies outside our own experience.

—Bonnie Bainbridge Cohen, *Sensing, Perceiving, and Action*

Perhaps the most sensuous experience is the feel of our own breath. That purr of life happening deep within the body links us to internal and external worlds. With each inhale, air becomes part of the internal body; with each exhale, it returns to the external environment, in constant exchange. The heart pumps oxygenated blood throughout the body so that every cell breathes. Voice is audible breath: the diaphragm determines volume, the vocal folds create the vibrations, cavities influence resonance, and the mind determines intention. As sound vibrates inner tissues of the body, we feel what we are saying, touched by the felt sensations of emotion. Sound resonates outer dimensions as well—other people, rooms, valleys, and concert halls. We touch them with our voices.

Breath

The process of breathing involves lung and cellular respiration. There are *five lobes* of the lungs, three on the right and two on the left. The *heart* in its pericardial sac is nestled between, slightly to the left. These contents—the lungs and heart—fill the container, the responsive rib cage. The right and left upper lungs are narrow at the top, extending up under the top ribs. Lungs broaden out toward the bottom lobes, which rest on the *diaphragm*. (Touch the floating ribs and feel their depth and width.) The muscular-tendinous diaphragm, essential in breathing, is dome-shaped, like an umbrella. It attaches around the circumference of the lower six ribs and onto the twelfth thoracic vertebra (in back) and the xyphoid process of the breastbone (in front). Thus, it makes an airtight container that seals the thoracic cavity, forming the floor for the lungs and heart and the ceiling for the stomach and liver.

Air enters the *mouth and nose* and travels down the *trachea*, the windpipe, to the bronchial tree (the respiratory units engaged in gaseous exchange). Midway behind the sternum, the trachea bifurcates into the *right and left primary bronchi* and branches again into the *lobar bronchi,* one for each of the five lobes of the lungs, terminating in *bronchioles*, where the exchange between air and blood occurs. Each *terminal bronchiole* further

In the Garden, handmade paper collage by Rosalyn Driscoll. Photograph by David Stansbury.

divides into multiple tiny *alveoli* sacs (air cells), where oxygen and carbon dioxide rapidly diffuse through capillary walls, based on pressure gradients. Oxygen-rich blood is pumped by the heart to all parts of the body, and carbon dioxide is expelled from the lungs. Three-fifths of lung volume is composed of blood and blood vessels, supporting this exchange. The familiar process of intake and release of air through the nose, mouth, and thorax is called *lung breathing.*

Organs of respiration. Anatomical illustration ©
2000 by Brian Hunter.

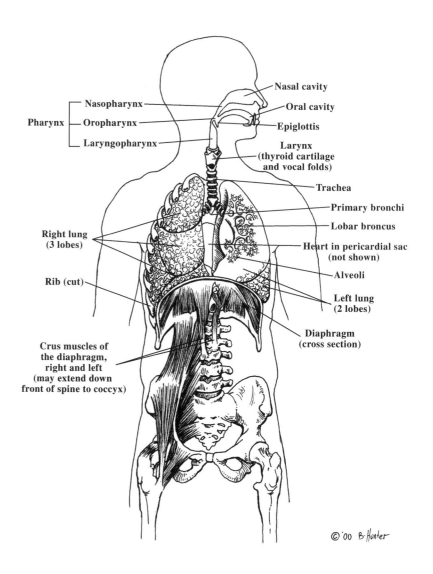

Nasal cavity

Oral cavity

Nasopharynx

Pharynx — Oropharynx

Laryngopharynx

Epiglottis

Larynx
(thyroid cartilage
and vocal folds)

Trachea

Primary bronchi

Lobar broncus

Right lung
(3 lobes)

Heart in pericardial sac
(not shown)

Alveoli

Rib (cut)

Left lung
(2 lobes)

Diaphragm
(cross section)

Crus muscles of
the diaphragm,
right and left
(may extend down
front of spine to coccyx)

© '00 B. Hunter

In resting position, the diaphragm creates an arched dome within the ribs. In contraction (the *in breath*), it descends down toward the pelvis, compressing the abdominal organs. On the release (*out breath*), the diaphragm returns to its domelike shape, expelling air from the lungs. Thus, the rhythmic contraction and release of the diaphragm massages the organs and assists in pumping fluids throughout the body. Two crus muscles, or crura, of the diaphragm attach to the front of the spinal column through connective tissue extending to the tip of the coccyx. Thus the tail and pelvic floor are involved in deep breathing.

The *medullary rhythmicity area* in the brain stem monitors the basic rhythm of respiration. To initiate breath, a signal from the nervous system must tell the crus muscles and diaphragm to contract. As the muscles shorten, they pull the diaphragm down (toward the lower spine) and the ribs flare open, expanding the thorax. The increase in rib space results in a decrease in interior air pressure in the five lobes of the lungs, and air rushes in through the nose and mouth to fill the vacuum. The out breath is a release of muscular contraction, allowing the diaphragm to return to its resting domelike position. In relaxed breathing, the in breath is work (lasting approximately two seconds), and the out breath is release (lasting approximately three seconds).[1] Breathing occurs three-dimensionally; the lungs and ribs expand front to back, side to side, and up to down, like blowing up a balloon. Any restriction at the ribs, waist, or abdomen can reduce breathing efficiency. Inefficient breathing results in lack of oxygen in the blood.

Expansion of the rib cage is assisted by the *external intercostal muscles,* and by the *scalene* muscles that suspend the top two ribs from the neck. These muscles are particularly noticeable in situations requiring increased oxygen intake (like after a fast run or when choking, sobbing, or sighing). Condensing of the rib cage is assisted by the *internal intercostal muscles,* which compress the ribs in coughing or forced expiration. Seventy-five percent of breathing is generally initiated by contraction of the crus muscles and the diaphragm descending; 25 percent, from the intercostal muscles and the ribs lifting.

Although breathing is normally monitored by autonomic centers, it can also be consciously initiated. We can easily "listen in on" the process of breath. Take three deep breaths, then return to your usual breath rhythm, observing the process. Thoughts and emotions are reflected instantaneously in the breath; one anxious moment can increase heart rate and tighten breathing muscles before we even know the thought has occurred. Try thinking of something that provokes anxiety in your life and notice the immediate physical results of breath. Try the same with a pleasurable image and enjoy the results. The ongoing exchange of interior and exterior landscapes happens reflexively, without our attention; it can also be observed and affected by conscious awareness.

Some people are "reverse breathers," lifting the shoulders up and flattening the belly to breathe in so that the diaphragm barely moves. Some people "overbreathe," working too hard to control the process on both in breath and out breath; and some of us are "shallow breathers," barely moving the bones and organs. (When we don't want to take in or to be seen, we hold our breath.) Breath patterns can reflect the rhythm of the mother's breath in pregnancy, the birth process itself, or later impressionable events in life that become fixed in the neural circuitry. Although there is no "right" way to breathe in all situations, it is useful to develop an attitude of open exchange with the environment, inside and outside. Tension restricts movement and masks sensation, numbing us from noticing the sensitive interplay of breath.

Cellular breathing is the absorption of oxygen and removal of waste materials that occur between each cell and its environment through the semipermeable cell membrane. As the heart pumps oxygenated blood, it travels away from the core through *arteries* to every part of the body and every tissue. Tiny *capillaries,* sometimes only one cell in diameter, exchange oxygen for waste materials through osmosis, according to pressure gradients, nourishing the cells. In this way, every cell breathes. Ninety percent of deoxygenated blood returns to the heart through *veins,* pumped by peripheral muscles and by smooth muscle fibers in the veins themselves. The lymph system filters about 10 percent of the blood before returning it to the veins. It takes

PRACTICING SOUND

Once, I spent several months in a private studio in northern California. Often, I would startle my host in the kitchen, because I moved so silently through space. Finally, she said, "Why don't you stomp around and make more noise, inhabit the space." I had spent so many years learning to be invisible, it was a curious invitation to take up space. I decided to practice making noise, letting myself be heard.

For years, I left voice workshops. I would begin the exercises, feeling the gentle flow of breath over lips. Adding audible sound for a hint of vibration, I would grow more adventurous. As a plethora of tones filled the room from the group, there would be a moment when I would become over-whelmed by the rawness of unmediated sound. I still can't find language to describe the sensation, its primitive power. But I would slip out the door, promising myself that, next time, I would stay.

Then I attended a workshop, led by Normi Noel, with my group of women writers. We began with our stories, each speaking briefly about the rose and the thorn in our lives. Then we spoke again, looking at each other as we spoke, breathing before, during, and after our words. This simple shift required that we feel what we spoke, and allowed the listener to feel with us. Speaking became theater, simply by allowing emotional impact. I thought how much less we would need to say if we felt our own words.

about three breath cycles for blood to travel through the body and return to the heart. Thus, the process of breathing includes both lung breathing and cellular respiration, nourishing and cleansing the body.

Voice

Voice is breath made audible. As air moves from the lungs through the trachea, it passes through the larynx, which houses the vocal folds. Mucosa-lined ligaments, or cords, run parallel to each other, from front to back of the larynx (from the thyroid cartilages to the arytenoid cartilages), forming a horizontal vocal diaphragm in the neck. If these folds are apart and relaxed, the passageway is open, and the air flows freely through from lungs to nose and mouth. If, however, there is an intent to make sound or speak, the folds are pulled taut, creating a thin space for the air to pass through, vibrating the cords. (If you have a cold, excess mucus irritates the cords, thickening them so that they vibrate more slowly, creating a lower voice.) These vibrations are amplified in resonating chambers of the body, shaped by the soft tissues of pharynx and mouth, and articulated by the lips, teeth, and tongue to produce audible sound. In speech, vowels vibrate and consonants interrupt, clip, and shape the flow of air. In sound production, the only muscles necessary beyond those of breathing are the tiny ones affecting the vocal chords. Tightening the superficial muscles of neck and shoulders, belly and back, only reduces resonance.[2]

Three components affect pitch: the size of the space between the cords, the tension of the cords, and the amount of air pressure that passes through the cords. The diaphragm controls the amount of air pressure; volume and timbre of sound are created by the various resonating cavities (like the pelvis, thorax, and skull sinuses), with different tissues having different resonant characteristics. The evolution of the larynx in humans changed the nature of sound production. In the throat, the separation of the pharynx and trachea (for breathing and vocalization) from the esophagus (for swallowing and digestion) was a risky undertaking because of the ease of choking. However, the advantages of vocal communication outweighed the risk and distinct structures evolved in humans to support the physical production and interpretation of sound and language.

The pharynx, normally considered the "throat," includes three distinct areas: nasopharynx (behind the nose), oropharynx (behind the mouth), and laryngopharynx (above the larynx). The pharynx continues down the front of the neck to the larynx ("voice box") and trachea. Vibrations resonate in the cavities of the pharynx and "color" sound. The pharyngeal cavities (nasal, oral, and laryngeal) comprise the primary resonating chambers for the overtones of pitches produced by the vocal cords. In *Sensing, Feeling, and Action,* Bonnie Bainbridge Cohen distinguishes the particular resonating characteristics of these regions: as we allow sound to resonate in the laryngopharynx, deep in the throat, it produces a self-reflective quality, vibratory and emotional. Sounds resonating in the oropharynx are more interpersonal—between self and other; and those in nasopharynx, mouth, and teeth are more abstract and objective in nature, distancing the sound from emotional coloration.[3]

Sound is vibration; we feel it in our bodies. Lower sensual and heart-stimulating pitches, in particular, vibrate in the large resonating cavities of pelvis and thorax; higher pitches resonate in the head sinuses. To separate feeling from voice, we have to restrict vibrations from moving down into the resonating cavities of the body. Yet as we dissociate our feelings from what we say, we create a dangerous split. If we can't speak what we feel, we repress feelings, imploding sensations back into ourselves. This contributes to ulcers, indigestion, heartburn, and many related problems associated with separation of feeling life—visceral response—from action and expression. In many ways, embodied voice supports the underpinnings of rational thought: ethical choice making, based on feelingful response to people and place.

Responsiveness of our full body through sound allows *resonance* with the places we inhabit. Resonance amplifies the sounds around us. The bird song, the rainstorm, the airplane overhead vibrate our tissues, or we wouldn't hear them. Removing fear from our bodies allows us to breathe deeply of the world: we take in, we release out. Varying breath rhythms brings up different states in the body, a process that has been explored for centuries through such forms as yoga, chanting, and martial arts. Breath is part of sounding, creating the power source for vocalization. Taking time to feel what we say reduces the necessity to speak, like taking time to taste reduces hunger and nourishes on many levels. Singing, toning, and vibrating the tissues of the body stimulates relaxation by encouraging blood flow, enhancing endocrine balance. Every moment of speaking, every word, can be a sensual, sensitive experience, full of life and learning and connection to the earth, as we reinhabit our voices.

આ   આ   આ

Lung breathing

5 minutes

Lying in constructive rest, eyes closed:
• Notice air coming in through your nose and mouth.
• Feel air passing down through your throat to your trachea, a long tube ringed with cartilage, supporting the front of your neck. Touch your neck lightly to feel this structure.
• Place your hands on your belly and allow it to expand on the in breath and condense on the out breath (without pushing).
• Imagine the shortening of the crus muscles, pulling the strong central tendon of the diaphragm down toward your lower spine.
• Place your hands on your ribs. Notice them fanning open on the in breath, softening on the out breath. Allow all twenty-four ribs to move as you breathe.
• Notice your breathing pattern. Enjoy a few deep, integrating breaths to finish your exploration.

When I heard of my father's death, I was dancing in a studio in Vermont. It was dusk, the last rehearsal of the day. My husband entered, gestured for me to come to the side, and said quickly, "Your father died at noon." I felt like I'd prepared all my life for that moment and was barely ready. I ran to tell someone, to hear my own voice saying the words: "My father died." And to feel my own response..

TO DO

Moving vibrations: Lips to pharynx*
5 minutes

Seated, eyes open or closed:
• Begin one sustained vowel sound, like "oooooh." Place it in the front of your mouth, vibrating your lips. Gradually move it back to the center of your mouth, opening the soft palate and tongue. Breathe whenever necessary and return to the same tone.
• Move the vibrations to the back of your throat (oropharynx), noticing the change in sound and the vibrations in other parts of your body. Move the sound down your neck so that it vibrates in your laryngopharynx and chest cavities; move it up the neck so that it vibrates in the nasopharynx and skull cavities.
• Now, in one long breath, bring the tone from the back of the throat to the center of the mouth, to the lips: feel the vibrations on the skin of your face.
• Try this sequence several times to become familiar with shifting your placement of sound.

*Exercise drawn from Susan Gallagher Borg, director of the Resonant Kinesiology Training Program.

Authentic Movement project. Photograph by Erik Borg.

Place visit: Resonance

Remaining in open attention, notice whether you can place sound in the environment around you, touching a tree, an animal, or the water. Consider this a conversation; what sounds does your place evoke? Try imitating the sounds of your place, creating a tone poem. Notice how this exercise connects to your speaking voice. Can you feel the resonance of place as you begin speaking? The in breath is active descent; the out breath is relaxation. The carbon dioxide released on your out breath is absorbed by plants; they, in turn, give off essential oxygen, in a constant exchange. 20 min. Write about your experience. 10 min.

CROSSING OVER: CEREMONY OF PURPOSE

My father roams from room to room looking for a place to be. He has been up since four, dressed as he was the night before in two soft shirts, a jacket, and a borrowed belt. Now he takes a dish and puts it in the oven, "Don't do that," my mother calls out. "Can't you just leave things alone!" Taking a teacup from his hands, she places it sharply on a countertop filled with muffins, tea bags, plates, tins, sugar packets, half-sliced melon—clutter I have known and cleared and fretted over since childhood. He stands, arms at his sides, and says quietly, "Can't you say anything nice to me?"

My father looks for empty spaces to put himself in along with the teacup. There are few to be found, even though the house has been emptied twice now, with tag sales and antique dealers, junk men and dish collectors taking things away. He has his own collection of objects under his bed, saved by announcing "not for sale" at opportune moments. I find them hidden from view: two hand-carved ducks, a camera bag stuffed with a clean shirt for quick departure, and black figure skates—blades sharp, ready for winter.

When my mother leaves the kitchen, my father washes dishes with too much soap. I notice him, minutes later, standing in the middle of the room, holding a milk carton, waiting for a connection. Placing my hand on his shoulder, I ask if I can have some. We share silence together as he pours milk into my apple juice, creating our own ceremony of purpose.

Read "Ceremony of Purpose" aloud, or write and speak your own story about breath and voice.

⚛ DAY 17

Air

In the world of modernity the air has indeed become the most taken-for-granted of phenomena. Although we imbibe it continually, we commonly fail to notice that there is anything there. We refer to the unseen depth between things—between people, or trees, or clouds—as mere empty space.
    —David Abram, "The Forgetting and Remembering of the Air," *The Spell of the Sensuous*

Air is the dynamic medium in which we live. The earth's atmosphere, essential to life as we know it, creates a thin envelope around the planet that affects our bodies at every moment. Although basically invisible, air provides the oxygen essential for respiration, screens harmful solar rays, provides the atmospheric pressure that keeps skin from literally exploding, and maintains the earth's temperature within a range suitable for life. Air also offers a transport medium through which scents, sound waves, water molecules, and many organisms travel. We smell, hear, taste, see, and feel in relation to the air around us as it touches our bodies and connects us to the world. Although we often think of sky as something above us, in fact we are standing in it, all the way to the soles of our feet.[1]

Air plays the body, like fingers on piano keys, stimulating a multitude of sensations, emotions, and images. The sight of a colorful rainbow, the smell of the ocean, and the sounds of gentle rain on a roof evoke visceral responses,

Off Route 22, Pawling N.Y. Photograph by Peter Schlessinger.

as do raging storms and wild winds. Shifts in light and dark elicit hormonal secretions from the pituitary and pineal glands. Phases of sun, stars, and moon mark time and place. Elegant patterns of ice crystals and snowflakes inspire awe and even humility in relation to the mathematical order inherent in the world. Breath itself is air, allowing the spoken word, song, prayer, and the abstraction of written language. Inspiration as well as spiritual and supernatural dimensions are the province of air, magical and ephemeral.

The composition of the atmosphere has changed several times in evolutionary history and continues to change today. During earliest formation, the atmosphere was primarily a mixture of hydrogen and helium gas held in by the earth's gravitational field. Over time, helium and hydrogen escaped by diffusion, and other gases were introduced into the atmosphere from continuous volcanic eruptions and collisions with meteorites. The photosynthetic processes of plants over many millions of years replaced much of the carbon dioxide in the air with oxygen. This change in atmospheric composition allowed all organisms that depend on oxygen for energy to evolve, including humans.

Current composition of the air we breathe, held within the thirty feet closest to the earth's crust, is nitrogen (78%), oxygen (21%), and argon (1%). Other elements are carbon dioxide (0.03%) and trace gases: methane, hydrogen, xenon, helium, ozone, neon, and krypton. Water vapor, as well as particulate matter such as dust and seasalt from the oceans, are the remaining constituents of today's lower atmosphere, making air visible. Humans are about midway in scale on the spectrum between molecules and storm systems![2] Yet we are changing the geophysical history of the planet by taking carbon from the ground and putting it in the air through incomplete combustion of fossil fuels in our cars and factories. Since atmospheric composition affects climate, we are indeed influencing the basic matrix that sustains our lives.

The atmospheric envelope that allows life on earth is a mere forty to fifty miles thick. The atmosphere as a whole can be divided into four distinct layers, characterized by differences of both temperature and density: the troposphere, stratosphere, mesosphere, and thermosphere. Humans interact primarily with the troposphere, which extends five to six miles above the earth's surface in polar regions and ten to eleven miles above earth in equatorial areas. The troposphere is the site of our weather patterns, and it contains about three quarters of the atmosphere's gases, in stable proportions. The top of the troposphere is a region called the tropopause, marking the transition to the stratosphere.

The second layer of the atmosphere, the stratosphere, begins approximately seven miles up and extends to about thirty-one miles. The stratosphere contains ozone that absorbs most of the sun's ultraviolet radiation, deadly to most organisms; it has little temperature change and few clouds. The cooler mesosphere extends to an altitude of about fifty miles, followed by the thermosphere, beginning above fifty miles and continuing to outer space; it is characterized by steadily increasing temperature with height. The different layers of the atmosphere rarely mix due to the sharp temperature

INVISIBLE AIR

Driving the forty-three miles between Middlebury and Burlington, Vermont, approximates traveling the depth of our atmosphere. This gaseous membrane, differentiating Earth from outer space, contains life-sustaining oxygen, swirling weather patterns, temperature-regulating clouds, and ozone layers. So much is held, mile by mile, it seems unbelievably precious. This short distance supports life as we know it.

Snow geese fill the Vermont skies in late October and again in April. Migrating from the Arctic tundra to their wintering grounds in the Carolinas or Gulf of Mexico, hundreds of thousands stop in cornfields preserved for their use. They prefer the same feeding grounds, mate for life, and live longer than some humans—sixty years or so. Blue and white, depending on age and species, they pattern the sky like clouds, their distinct calls inviting us to look up in awe at the precision of their V-shaped patterns and the strength of their wind-born wings. When I took my class to the Dead Creek Wildlife Refuge to experience this massive display, study the uniqueness of individual birds, and note their reliance and effect on habitat, one student said: "The only birds I knew in Manhattan were pigeons in the park. I didn't spend much time looking up at the sky. It wasn't safe."

breaks that define them. Life-supporting ozone occurs in the stratosphere, yet ozone also exists in the troposphere, where it traps sunlight and heat reflected from the earth's surface inside the atmosphere, increasing temperatures and creating what is called the greenhouse effect.

Weather can be viewed as the immediate state of the air in any one place. As it swirls around the globe, air responds to the rotation and topography of the earth and the heat of the sun. Because the atmosphere is a medium of connection, change in one area has distant as well as local effects. Air is moved around the earth by six belts of prevailing winds that vary according to latitude. As the earth rotates on its axis, the air can go slower or faster than the ground, affecting weather patterns. The consistent interactions of temperature, wind, humidity, latitude, altitude, ocean currents, and topography result in large climatic regions called biomes, characterized by different plants and animals. (Tundra, boreal forests, deciduous forests, tropical rain forests, grasslands, and deserts are regions that reflect long-term effects of weather patterns.) Wind circling the globe redistributes the earth's surface, dissipates pollution, and transports seeds, a visible reminder of the movement patterns of air.

Air also moves vertically from one layer to another. Air is warmed as energy from the sun enters the atmosphere as sunlight and is reflected back from the earth's surface. Warm air has fast-moving particles and is thus lighter, creating low pressure on the surface below it. Cooled air has slower moving particles and is contracted and denser, creating high pressure. Uneven heating causes currents that move the light, warm air up and the dense, cold air down. Winds occur when cool air flows in to fill the area the rising warm air has vacated. The greater the difference in pressure between the areas, the stronger the winds will be. Vertical change in air also affects precipitation. Because temperatures are generally cooler at higher elevations, when the warm air that carries latent heat in the form of water vapor rises, the water vapor condenses and falls as rain, hail, or snow. Thus, clouds usually drop their rainfall on the side of the mountain facing prevailing winds, creating situations in which one face is lush while the other is arid.

Particles scatter sunlight, creating familiar blue skies. Without particulate matter in the air, the sky would be black except for the sun, stars, and reflections from planets. Small, light-scattering particles suspended in the air can include rock fragments moved by dust and winds, salt spray from oceans, and smoke from forest fires. When sunlight is scattered by these airborne particles, the blue spectrum of light (short wavelengths) dominates the view, and we see the sky as clear blue. The sky is less blue when there is dust in the air, and whiter when there are clouds in the upper troposphere or sulfite aerosols in the air. A fuller color spectrum is made visible in rainbows, formed as sunlight is reflected from falling precipitation. When the sun nears the horizon, light-scattering particles create a greater proportion of red light (long wavelengths) than blue, producing sunsets.[3]

Particles in the air vary in size. Ash from large forest fires, spider webs, soil particles, pollen grains, living microorganisms, and large airborne debris from disturbances such as tornadoes or volcanic eruptions are present in the troposphere. Fine particles can remain in the air for weeks or even years, while most of the denser, heavier particles generally remain in the air for a few hours or minutes before falling to the earth. Particles suspended for over an hour begin to affect atmospheric reactions; they can also interfere with sight by obscuring, refracting, or reflecting sunlight or moonlight, as in the long-term effects of ash after volcanic eruptions or massive fires.[4]

Clouds clean the air. Formed by water droplets and/or ice crystals of various sizes, gathered around particles, they circulate around the globe, removing foreign matter from the skies and returning it to the earth through precipitation. Clouds vary in shape, occur in different layers in the troposphere, move at different speeds and in various directions, and change moment by moment. Cloud formations and atmospheric pressure are ready indicators of weather, and they are essential for farmers, navigators, travelers, and others dependent on predicting changes in the environment.

Cumulus clouds are somewhat flat-based, puffy white, and heaped water-droplet clouds, generally floating through the sky in fair weather at various levels. *Stratus* clouds are gray, water-droplet clouds, low and usually seen in calm conditions; when they near the ground they may become fog. *Cirrus* are thin and wispy ice crystal clouds, high in the sky (five miles or more). *Nimbus* are rain-producing and generally dark gray. Most skies produce a combination of cloud types: *stratocumulus* clouds involve both heaps and layers, and *cumulonimbus* (thunderheads) are piled high in various shades of gray. Cumulonimbus are warning clouds, foretelling cloudbursts, hail, thunderstorms, tornadoes, or waterspouts (like tornadoes but over water). If you are out boating when these appear, head for shore![5]

The sky is full of electrical potential. Electrical charge produced by storms creates lightning, "an electrical discharge between one part of a cloud and another, between two clouds, and between clouds and the earth."[6] To determine the distance of a lightning strike from your location, by miles, count the number of seconds between the sight of lightning and the sound of thunder. If they are almost simultaneous, the lightning is close! Tropical storms such as hurricanes (with winds over 73 mph) involve massive release of atmospheric potential energy, forming a spiral pattern around the eye, or vortex, lifting heat off the earth for dispersal while causing considerable destruction to land and waters below.

Air pressure exerts force on the surface of the earth and our bodies—14.7 pounds of force per square inch at sea level. Changes in air pressure are monitored carefully in our structure to balance inner and outer environments, especially by the inner ear. In fact, compression and weight on bones stimulates bone growth. Without the effects of air pressure and gravitational pull on their bodies, astronauts in space find that their bones begin to decalcify. Although air is a less dense medium than the water of our origins, the human body is still supported and affected by the gaseous matrix in which we live.

The sky can be easily obscured in our contemporary lives. The all too familiar brown or whitish haze over cities is a telltale signal. Tall buildings also can block the sky; stray light from highway lamps, billboard signs, and mall parking lots make sky viewing at night difficult. In fact, sites providing

When I run, I feel air inside and out: under armpits, between legs, and through nostrils. It is a partner in the process, the medium through which I move. Fourteen pounds of pressure per square inch of skin informs the tone of my muscles, the density of my bones, and the reach of each stride. Breath remains constant, regardless of dynamics, so the body can continue its full range of expression without depletion. So vast, yet almost invisible, air is essential to lungs, blood, and cells, just as it is to the wood, water, and soil around me.

Yoga: Sun Salute, Surya Namaskar.

a clear view of the sky (in day or night) are now sufficiently rare that sky-viewing parks are being created in out-of-the-way regions such as northern Canada. When clean air and minimal light emanate from earth, the night sky reveals magnificent displays of stars, constellations, and our own Milky Way galaxy visible to the human eye. Creating these parks involves long-range vision, such as agreement from surrounding towns to restrict development and night use of electrical lights.

The sun, moon, and stars are familiar landmarks, helping us to orient in time and space. Eight other planets and their moons make up our solar system, orbiting in roughly the same plane around the sun. Venus is usually the brightest planet in the sky, alternating between morning star and evening star, depending on its orbit phase. Orange-red Mars is the closest planet to us, rising at sunset and remaining in the sky through the night. Meteors, or shooting stars, are tiny chips, from sand to marble size, knocked off tiny planets called asteroids or blown off comets, which burn up as they strike our atmosphere. In North America, annual meteor showers peak in mid–August and mid–December. The best time to view the night sky is on a clear night, after rain so that pollution has been washed from the air.[7] Sky gazing brings the depth of space into our bodies, where it is matched with equally vast dimensions.

Interconnectedness is clearly represented in our relationship to air. Once air is polluted, it will pollute all other substances on the earth. Pollutants released into the air not only travel on air currents around the globe, they fall into the soil and the waters through the process of gravitation or sedimentation. Toxins become part of our drinking water and, eventually, of our bodies. Climate, seasons, and the sun, moon, and stars affect our human moods and imagination, as well as our physical bodies—our environment. Although we think of air as invisible, in many ways, it is the medium that links us with each other and with the earth itself.

TO DO

Flute*

10 minutes

Standing in postural alignment, eyes closed:
• Bring your attention to the senses of your body. Take time to feel impulses and internal sensations. Notice your ongoing interaction with the movement of the earth and the specific environment around you.
• Now bring your focus to the ground. Imagine a flute playing music into your feet. Let the air pass up through your legs and pelvis, spine, ribs and arms, neck, and skull. Allow yourself to be moved by the air, the rhythms, tones, and melody of the flute, playing through your feet, into your body.
• Remember that a flute relies on breath rhythms, with starts and stops, fluctuations in sound, and silence. Allow these subtle changes to resonate in your body. 10 minutes.

*Exercise by Felice Wolfzahn, dancer and teacher of Contact Improvisation.

Empty spaces (alone or with a group)
15 minutes

Standing, eyes open, in a big room or outdoors:
• Begin walking in any direction, becoming familiar with the parameters of this space or place.
• Continue walking at a fairly quick pace, weaving in and out of the empty spaces around you. Move between, over, and under other bodies or landmarks (trees, rocks, and shrubs).
• Keep walking and increase the speed even more so that your reactions occur more quickly, automatically. Continue to find empty spaces to move through. Notice how you feel when you are close to someone/something or far away. Where does proximity register in your body?
• Begin to run, dodging in and out of the spaces as they appear. Use both the center and the periphery of your space.
• Return to a fast walk, moving in and out of the empty spaces. Notice how much more "room" there is when everyone is aware of space.
• Return to a normal walk, noticing the air around you.
   Pause. Close your eyes. Notice postural sway and the inner dance of the body in stillness. Feel or imagine the spaces inside your body.
• Open your eyes, inclusively attending to the space inside, the space outside, and the air that connects them.

Photograph by Erik Borg.

Place visit: Attention to air

Visit your place at night; feel the orientation of moon, stars, and planets. In what phase is the moon, and where is the North Star? Learn one constellation to accompany you on future visits. Each season brings a different view. For example, from New England in fall, you might see more distant stars, like the Andromeda Galaxy. In winter, as Earth's night side faces a different direction, you would look through the spiral of our disk-shaped Milky Way galaxy, with vibrant, young stars visible to the eye. And in spring the Big and Small Dippers are clear, with the two stars at the end of the Big Dipper's bowl pointing almost directly at Polaris, the North Star. In summer, Earth's night side faces the luminous center of the Milky Way, so you might see a hazy band of light.[8] How is your relationship to place changed when you visit at night? 20 min. Write about your experience. 10 min.

CROSSING OVER: CALLING

I call my father when I want to talk to someone who sees through things. There is always a telling line amid the jumble. "It's all lies and promises here," he responds when I ask how things are going. Then he adds, "Can't you just take me out and shoot me?"

My father has hidden the portable phone somewhere in the house; no one can find it. "Why would he do that?" my mother asks. When I talk to him, he tells me: "You just have to keep something gazing through air."

This month he goes into a nursing home. He hasn't been told, but he knows. He sits now at my mother's elbow, wherever she is. He sees them gathering his shirts, underwear, shorts. I call each night to hear the news. Our neighbor in Maine describes it simply, "It's like cutting a flower and putting it in a vase." At dawn he waits in the car, dressed for departure. When my sister comes searching, he says simply, "I'm going to miss it here."

I collect my father's words on little pieces of paper; tuck them in pockets, notebooks, and purses. I find myself repeating them, as if to learn a language long forgotten yet familiar. When I speak these sentences from memory, I lose their gentle twist. It is a language developed from years of seeing several worlds at once and keeping quiet about it.

Read aloud or write and read your own story about air.

DAY 18

Muscle

We live in our animal body.
—Terry Tempest Williams, Breadloaf Writers' Conference, 1997

Muscles move us through the landscape. Skeletal muscles activate our bones, smooth muscles stimulate motion of organs and blood vessels, and cardiac muscles pulse the heart. All provide internal feedback about the body and external connection to place. Over six hundred skeletal muscles work together as an interconnected whole to protect other tissues, offer tensile support to our bones, and help pump vital fluids that can get stagnant from inactivity. They also provide our capacity for expressive movement, spanning a continuum from the subtle warning of a raised eyebrow to the stormy gesticulations of a fight.

BODY LANGUAGE

My anatomy and kinesiology professor taught about gear-rationing of muscles by imitating a squirrel: super-fast, freeze, and an easeful lope across the grass. He stood in front of the class, twitching like a familiar gray squirrel, while everyone was laughing. Then he contrasted this demonstration with the qualitative range of human movement. From delicate to wild, the same person can stroke a hair on a baby's cheek and rage in anger. What's unique to humans are the smooth transitions, the speed and seamlessness of change, and the expressive possibilities we move through each day. When I teach, I can communicate mood and intent through my muscles, without saying a word.

Set for *Crossing Over*, by Herb Ferris. Laminated ash. Dancers, Paul Matteson and Lisa Gonzales. Photograph by Bob Handelman.

When I teach the names and theory of muscles, students become irritable. It is humorously predictable. As we study the bones, everyone is calm, direct, and clear. But muscles bring up heat, aggression, a need for action and interaction. Students argue about test scores, are grumpy about too much work, and get embroiled in lengthy discussions. This is familiar: Western culture is muscle-oriented; action is a priority. Once I point out the difference, the whole class understands. We can go back and forth, shifting our attention from bone to muscle, noticing the change. Working with nerves and bones leads to new patterns. Working with muscle reminds us of our habits and also of our resistance to change.

Muscles are the meat of the body. Comprising about 70–85 percent of the body's weight, they link us to our animal relatives through nutritional requirements and physical characteristics. Muscles use most of the food and oxygen we consume and contribute to the contours and coordinations of our form. Just as bones need a certain amount of compression to stimulate their health, muscles need movement to remain toned and responsive and to flush toxins. Maintaining balanced tone requires a relationship between strength and stretch in all the muscles of the body, preventing joint stress. Too much muscular tension creates a condition of hypertonicity (you can't sit still) and too little creates hypotonicity (you can't get moving). In general, movement creates a desire for more movement; when muscles are toned, they want to move.

Muscles must have a signal from the nervous system to contract. The fastest neuromuscular path is a reflex arc, making an efficient loop from a sensory nerve in the skin (or deeper tissue) to the spinal cord and returning through a motor nerve to signal a muscular response. Simultaneously, neurons within the spinal cord (interneurons) inform the brain about what is occurring. In other words, you can withdraw your hand from a hot stove, before the pain or action even registers in the conscious mind. Many movement reflexes are encoded from our evolutionary heritage, providing a foundation for the demands of an upright stance and more complex behaviors. The reflex arc in the lower legs for postural sway keeps us from falling; as we sway forward over our toes, the calf muscles (soleus) contract; when our weight is pulled back to our heels, the muscles release and the process repeats. When there is stimulation to the bottom of our feet, they push (flexor extension pattern), instantly triggering whole-body coordinations necessary for standing.

Although generally thought of as a motor system, skeletal muscles also stimulate and record sensory information. Moving our torsos and limbs generally feels good, but muscles can also be a repository for stress, registering soreness, stiffness, and generalized tension. Proprioceptors (self-receivers) are specialized movement receptors that tell us about position and motion in space of every body part. They include the muscle spindles and golgi tendon organs (measuring force and pull on muscles and tendons), joint receptors (registering compression), and the maculae and cristae of the inner ear (monitoring location of the head in relation to gravity). Proprioceptors also monitor the amount of force necessary for effective action and interaction. Movement memory (kinesthesia), stored in the cerebellum, can be triggered into consciousness by a partial movement. You might recall the precise position of your body when you experienced your first kiss or received important news, by moving your head or hand in a particular way.

The muscular system, in fact, memorizes sequenced movement patterns based on past sensations. Established in the somatosensory cortex, these sensory engrams are based on your unique life experiences and inform ongoing relationships both to your body and to the earth.[1] As you become skilled or habituated in any activity, the sensory engram matures and becomes refined, and eventually the activity happens "unconsciously." In this process, from the awkward first attempts through refinement and habituation, the skill eventually becomes automatic, allowing the mover the freedom to focus beyond the task at hand. For a dancer or skier, this might mean the capacity to concentrate on the expressive qualities of performance (during a difficult turn or balance), because the technical skills are inherent.

Some engrams, such as those established through childhood trauma or more recent physical or emotional distress, may result in practiced, ingrained behaviors that are detrimental. For example, a child raised by an alcoholic parent may, as an adult, hold back from expression or have a retreated chest because of fear of being attacked. This learned behavior, once effective, may now restrict communication or hurt the spine. As we work with muscles to explore new movement patterns and sensations, we may uncover engrams that are limiting as well as those that are effective. Postural habits can "feel right" even though they are detrimental; it takes time to learn new sensory engrams and incorporate them into our lives.

The relationship between the nervous and muscular systems explains the effectiveness of visualization techniques and imagery in changing muscle patterns. Muscles respond to signals from the nervous system based on the need to move rather than on specific muscle directives (bend your arm, rather than contract your biceps muscle). They also work as a coordinated group rather than as individuals. If someone directs you to "relax your trapezius muscle" to initiate shoulder release, very little will happen. But if the facilitator uses an image, like "imagine your arms are tassels dangling from your shoulder sockets," you may feel a change. Or if you are encouraged to lie down, close your eyes, and visualize yourself resting on a beach as both shoulders melt into the sand, you may engage visual pathways in the brain to initiate a relaxation response. (Unless you hate beaches!) Many athletic trainers, dancers, and physical therapists use imagery and visualization techniques to refine muscular habits, engaging more efficient patterns.

In determining the function of a muscle, you must first know where it is attached to bone and which joint(s) it crosses. Muscle attaches to bone through tendons and by weaving directly into the outer covering (periosteum) of bone. Since the basic role of muscle is to move bone, there will be an attachment site on at least two bones, crossing one or more joints. For example, the rectus femoris of the quadriceps, the central thigh muscle, attaches on the front of the ilium (pelvis) and on the front of the tibia (lower leg), spanning both the hip joint and the knee joint. The rectus femoris can both flex (decrease the angle of) the hip joint and extend (increase the angle up to 180 degrees of) the knee. It is therefore called a two-joint muscle.

In standing anatomical posture (arms to the sides, palms forward), the attachment site closest to the center of the body through the skeletal structure (not through space) is called the origin of the muscle, the one farthest from the center of the body is called the insertion. In a moving body, one bone is generally stable and the other is moving, in which case the stable end is called the origin, and the moving end is the insertion. Thus, when working with the rectus femoris in standing posture and lifting one leg, the origin is on the pelvis and the insertion is on the tibia. Bending forward from the hip,

When I attended the Breadloaf Writers' conference, I took along for review a series of stories about my father's illness with Alzheimer's disease. In our small-group sessions we read and critiqued each other's work. My discussion focused on form; the professional writer who facilitated described my writings as vignettes, neither short story nor novel, and suggested I was leading toward a novel. I understood his perspective, began planning a larger view, and felt my neck muscles tighten. During the week, my right arm went into spasm. I could not turn my head and left the conference early due to the pain. As I drove over the mountain, all the discomfort subsided. I started to laugh: my right arm, given its freedom, would have punched him out. As an artist, I know the creative process well, the readiness or resistance of content to manifest into form. My writing wasn't ready, and my body protected what I wasn't ready to say.

however, the origin becomes the stable tibia and the insertion is the moving pelvis. It is also useful to determine the fiber direction. Many muscle fibers run the length of the muscle between origin and insertion, with parallel fibers; but some are pennant- or fan-shaped, which exerts a different pull on the bone. Because no muscle works alone in a joint action, there will be muscle fibers exerting pull in various directions.

A primary initiator of a movement is called an agonist (there may be more than one). In our example, the rectus femoris is an agonist in the process of hip flexion. The paired muscle is rather unfortunately called the antagonist, although the two cooperate to effect an action. The antagonist is on the opposite side of the body from the agonist and crosses the same joint, with similar origin and insertion points. The antagonist of a paired muscle generally releases (lengthens) as the agonist contracts; sometimes it will contract as well, to stabilize a joint or to offer resistance to the movement (for more controlled action). In our example the antagonist of the rectus femoris is the hamstring muscle, which has its origin on the bottom of the ilium (the sit bone) and its insertion on the back of the tibia and fibula (lower leg). Thus, it crosses the same joints as the rectus femoris and creates the opposite actions: hip extension and knee flexion. Balanced tone (both strength and stretch) in paired muscles is essential for efficient joint alignment at both the hip and the knee. Muscles stabilizing the joint are termed fixator muscles; this is a relative term, determined by function. When standing and swinging one leg, muscles of the opposite leg stabilize the pelvis (as in walking); switching legs, the muscles change roles.

Muscles pull rather than push. Collections of muscle fibers (cells) comprise the belly of a muscle, the site of muscle contraction. Sarcomeres are microscopic contractile units arranged into myofibrils running the length of each muscle cell. When a nerve impulse is sufficient to trigger a muscle to contract, the various sarcomeres draw together (within the length of a myofibril, within a muscle fiber, within a muscle bundle, within a muscle unit, within a muscle). This shortening causes the muscle to contract toward its stable bone end, or if both bones move, toward the muscle's center. Each cell contracts to its maximum; the number of muscle units (fibers innervated by the same motor nerve) engaged constitutes the strength of action. Movement occurs on a spectrum from light to forceful, highly refined to bold, depending on the number of fibers called into action and the ratio of nerve to muscle cells within a motor unit. Face muscles, for example, have a ratio of 1:10, allowing varied facial expression; the large gluteus maximus of the hip has a nerve to muscle ratio of 1:1,000 or more.[2]

Smooth connective tissue (fascia) covers muscle at every layer, allowing the components to slide over each other during contraction and offering support for vessels and nerves. The merging of these fibrous layers forms tendons at the ends of muscles for attachment to bones or other fascia. For a reminder of muscle structure and its relationship to fascia, take a look at steak or chicken in your local supermarket. The parallel fibers of meat are the muscle, the white and transparent tissues are the fascial coverings.

Muscle tissue has the unique ability to contract and release or more specifically, to shorten, lengthen, or lock into place. Hardening or softening of the tissue can occur almost instantaneously once a nerve impulse is sufficient to trigger a response. The three possible contraction responses for a muscle are (1) a *concentric* contraction: a shortening contraction in which the muscle contracts (shortens), and the bones move toward each other (example: bringing your torso off the floor and toward your legs in a sit-up); (2) an *eccentric contraction:* a lengthening contraction in which the muscle contracts *as it is being lengthened* by an outside force or gravity, and the bones move away from each other (example: lowering your torso to the floor with control, away from your legs, from a sit-up); and (3) an *isometric contraction:* a holding action in which the muscle contracts but no movement occurs because the opposing forces are equal (example: holding halfway up, in a sit-up). Isometric and concentric contractions generally build muscle strength and bulk; eccentric contractions develop strength and length. (See p. 133.)

Although we are born with a certain number of muscle fibers, we rarely, if ever, use our maximum potential. Each fiber must be vascularized to be brought into action. Muscles respond to the demands placed upon them, creating a highly responsive system. For example, if you learn a new activity that requires increased flexibility, the system responds by changing the resting length of appropriate muscle groups. The activities we do and the sensory engrams we have built literally shape who we are. Our muscle mass and responsiveness affect the way we think of ourselves, and the way we think affects our muscles and our interactions with the environment. Through neuromuscular plasticity, our capacity for change, we can influence the way we move, open to new sensory engrams, and explore new thoughts about what it means to be human in relation to the world.

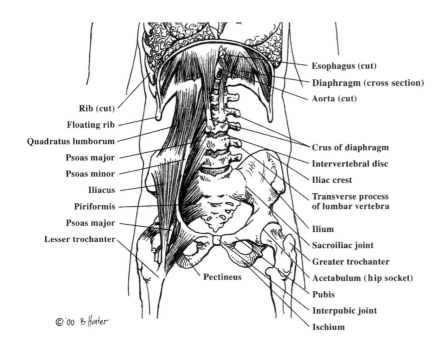

Anterior view of iliopsoas and deep muscles of the trunk. Anatomical illustration © 2000 by Brian Hunter.

OF SPECIAL INTEREST:
THE GRACILIS MUSCLE

Gracilis, or "Gracie" for short, is a long, thin, two-joint muscle connecting the pubic bone of the pelvis with the bottom of the knee. It draws the leg inward, toward the midline. When tight, it can also create havoc with pelvic alignment, causing severe strain on the back or knees. In the past several years, I have worked with five different women scheduled for surgery due to spinal pain—to insert a metal rod or fuse vertebrae. The focus in our sessions was an exercise called hip circles, which gently moves the leg bone in its pelvic socket, stimulating the joint capsule. Sometimes the leg won't move at all; other times there are tiny twitches, progressing to trembling, livelier twitching, and kicking. As each woman lay on her back, opened her knees, and gently massaged this muscle on the inner leg, it felt like a steel cord with swollen, pea-shaped collections of fluid, signifying tension. In each case, working simply, slowly, gently with hip circles, the gracilis muscles released their hold, and the woman was spared invasive surgery on her spine. Although there are situations where operations are essential, first listen to the body's call for attention.

For those of us who hide our tension in our hips (where no one can see it), who sit with tightly crossed legs (so proper!), who have a history of intimidation or abuse (verbal or physical), or who experience stress (who doesn't?), this joint can become immobilized. As we turn or twist, immobility at the hip transfers movement up to the sacrum or down to the knee joints, which are ill-equipped to accommodate increased range. Hip circles feed neurological input into the hip joint and remind the brain of the full movement potential of the joint. If you have back or knee problems, I tell my students, it's worth checking in with Gracie to see how she's doing.

The main function of muscle is to move the body, not to hold it together. If the bones are vertically aligned, providing space for soft tissues, ligaments and tendons can stabilize joints, while balanced muscle tone links the various parts of our dynamic bony scaffolding. Three deep postural muscles support the central core of our upright stance and underlie efficient functioning of other muscles: the *longus capitis and colli* (integrating skull to the front of the spine), the *iliopsoas* (integrating the front of the spine to pelvis to legs), and the *soleus* (integrating the back of the lower leg to the heel). These muscles work with the crus muscles of the diaphragm to allow the rest of our musculature to be ready to move with subtlety or force as required.

To locate the basic postural muscles more specifically, begin with the longus capitis and longus colli muscle group. These muscles connect your skull (in front of the occipital foramen, behind your throat) to the body of each vertebra of your neck, down the front of your spine, attaching to the first, second, and third thoracic vertebrae (T-1,2,3). Initiate the "yes" nod and roll your head forward, vertebra by vertebra (tucking your chin), toward the front surface of your spine. In standing postural alignment, support from the longus capitis and colli muscles allows the large neck and shoulder muscles to relax. The muscles suspending your ribs and shoulder girdle from your head and neck (scalenes and levator scapulae muscles), in particular, should remain supple and resilient (rather than gripping your head onto your body).[3]

The iliopsoas muscle group lies along the inner surface of the lower half of the spine, originating at the twelfth thoracic vertebra (T-12) and attaching down the front of each lumbar vertebra (behind the organs). The iliopsoas major merges with fibers from the iliacus muscle inside the ilium of the pelvis on each side, then travels over the pubic rim and inserts on the inside top of each leg, the lesser trochanters. Thus, in postural alignment, the iliopsoas integrates spine to pelvis to leg. The soleus muscles connect the backs of your shinbones to your heels, contracting and releasing to keep you from falling as you sway forward and backward over your feet. In efficient alignment the three basic postural muscle groups keep you upright. The big surface muscles on the back of your neck and shoulders, spine, and hips, as well as the front surface of belly and thighs, can relax.

Muscles provide a sense of self as a responsive being, emphasizing our efficient animal nature. Exploring three-dimensional movement, such as playing frisbee, dancing, or interacting with children or animals, opens unusual patterns and gets us out of habituated movements. Taking regular breaks in our workday to change position, move around, and travel outdoors can help reduce muscle fatigue and overuse syndromes (injuries) from repetitive muscle actions. Walking, the basic human striding gait for which our muscles are exquisitely suited, offers a simple and cost-free way to invite whole-body fluidity. In general, for healthy muscles and mind, we can build diversity into our jobs and our lives, allow patterns of activity and rest, engage both intense focus and play, communicate through our whole body with the world around us, and respond to change with inherent adaptability.

Duet from *Felt Presence of an Absence*, choreography by Peter Schmitz. Dancers, Sean Hoskins and Paul Matteson. Photograph by Bob Handelman.

Strengthening and lengthening: Concentric, eccentric, isometric
5 minutes

To identify and strengthen the basic postural muscles, you can engage the three types of muscular contraction: concentric, eccentric and isometric. Lying on the floor in constructive rest, eyes closed:

• Begin the forward action of the "yes" nod, bringing your chin toward your neck. Roll, vertebra by vertebra, raising the head and neck until the ribs begin to come off the floor. You are initiating a sit-up action by *concentric* contraction (shortening) of the longus capitis and longus colli muscles of your neck. (If you can articulate each vertebra, you are using the right muscles. If the neck moves as one piece, however, you are gripping your larger muscles.)

• Place your elbows on the floor by your sides; roll each lumbar vertebra, one by one, off the floor, hollowing the belly toward the spine (as though you are lying in a hammock) until you have rolled your torso all the way up

TO DO

over the sacrum and coccyx. (If you can move each vertebra, you are engaging the iliopsoas muscle group in concentric contraction. If your spine moves as a whole, you have jumped to your abdominal muscles on the belly surface; slow down and try again.)

• Hang forward over your legs in seated position, knees open to the sides, feet together, and breathe deeply.

• Now return the legs to parallel, and reverse the action. Lower the torso slowly, with control, toward gravity, using an *eccentric* (lengthening) contraction of the iliopsoas and abdominal muscles. The muscles are contracting, but they are being lengthened as they lower toward gravity.

• Pause halfway in stillness. This pause requires an *isometric* contraction: the muscles are contracting, but there is no movement. (Your muscles may shake trying to hold in this position.)

• Continue rolling down the spine, and lower the neck slowly, with control, vertebra by vertebra, an eccentric (lengthening) contraction of the iliopsoas, longus colli and longus capitis muscles.

• Then transition to standing. In postural alignment, allow your weight (plumb line) to fall forward over your toes until the soleus muscles on the backs of your lower legs begin to contract (keep your heels on the floor). Feel the plumb line return back toward your heels.

This action, postural sway, constantly engages the soleus muscles in concentric contraction and release, keeping us upright in standing alignment. Notice what happens to your breath and to your mind when you are off balance and when you are on balance.

Gear-ratioing of muscles (alone or in a group)
10 minutes

Begin walking, eyes open:

• Call out a word that registers an emotional state, connecting language to movement. Don't "think too hard" when choosing a word; say whatever comes to your mind that implies motion.

• Say one word at a time, however bold or subtle, with plenty of time between to let the movement register in the body as, for instance, funny, deliberate, mean, happy, angry, ambivalent, tender, wild, ecstatic, stupid.

• As you (or someone else) calls out a word, allow your movement to change to that quality, whatever it is to you. Keep moving, and take time to let all the muscles of your body gear-ratio or find the appropriate level of tension to register the state. It may not look like anyone else, or like your idea of the word; let the muscles tell you their response. After all, muscles do this exercise all day long!

• Call out another word; let the body respond. Gradually increase the speed of the words to surprise yourself; notice how quickly you can respond. If you lose an authentic connection, slow down, remaining nonjudgmental.

• Now change your use of language and call out a poetic phrase to evoke a response beginning with the word "like"—like rain on water, like green air, like floating on grass, like wild wind. It doesn't have to make sense. Allow your body to respond to these images in movement. Images are the language of the unconscious, evoking association and connection. The movement will be unique, personal to you; let yourself be surprised by what comes. Try to keep moving even as the words are being spoken so that one muscular state or engram changes fluidly to another. Notice how your muscles feel as you move, engaging their expressive capacities. Trick yourself into moving before you can think, judge, or censor.

• Pause, in open attention. Begin walking. Imagine all the potential in your muscles for expression as you walk.

Lion
5 minutes

Seated comfortably on a cushion or the floor, legs crossed:
• Place your hands in front of you on the floor for support, like the forelegs of a lion or lioness.
• Open your eyes and mouth as wide as you can, and stick out your tongue, stretching the inside surfaces of your mouth. Roll your eyes upward in their sockets.
• Relax, and repeat several times. Pause in open attention.
• Try it again, bringing your hands in front of your chest, like claws. Breathe in, and on the out breath make a deep sound from the belly, roll your eyes upward, and stick out your tongue.

This yoga exercise activates the muscles of the face, eyes, and tongue; it also reminds us of our wildness.

Place visit: Light touch duet*

Standing in efficient alignment, feeling the support of your postural muscles and allowing the rest of your muscular system to release: Do a body scan to notice unnecessary places of tension and holding. Then try a light touch duet at your place. Close your eyes and notice the first sensation of touch. You might feel yourself being touched by the wind and responding in movement; being touched by the song of a bird and responding; being touched by the texture of leaves under your feet and responding; being touched by the scent of the soil and responding. Add vision and continue. There is no right or wrong; follow the muscular reactions as they occur. 20 min. Write about your experience. 10 min.

*Exercise drawn from Felice Wolfzahn, performer and teacher of Contact Improvisation.

CROSSING OVER: BRIDGE

My father walks and walks. Each day he goes to the park, traveling a precise route so that he doesn't get lost. When we visit, he invites me to join. We cross the road to the oak ridge, stroll down the hill to the duck pond, up the ravine to the solitary pine, and then to the footbridge. As he points out the way with a stick, I see each place has its purpose. On the hill, he tells me to press my spine against the rough trunk of a hemlock. And from that unusual view, all the oaks dissolve into one straight line, planted years before. As we arrive at the duck pond, swans and mallards greet us for their daily feed. The lone pine tree on the slope harbors a bird's nest with two speckled eggs. At the bridge the runners' path narrows across an old train trestle spanning a road below. We walk to the center, and he turns around. "We can't go any further," he admonishes. "We might not get back." I understand: he is practicing crossing over. Someday, he will not return.

Read aloud, or write and read your own story about muscles.

❧ DAY 19

Animals

COMPANY

Snow reveals animals. Deer and the occasional moose leave telling prints in deep drifts. The single-file track of a coyote crisscrosses those of a cottontail rabbit. An air hole, rimmed by ice, denotes breath below (the possible den of an opossum, remaining indoors until night). Otter slides link one mound of snow on the riverbank to the next. A flock of wild turkeys leaves three-pronged claw marks, plentiful and distinctive. Drag marks lead us to a skinned beaver in the woods; tail, teeth, claws, and fibrous muscle left as hunter's bait to lure a bobcat. Bear scratches mark a nearby tree. One hundred fifty years ago most of these animals were "extinct" in this region of New England. Now, when students at the college tell me there are no animals to view, I remind them to wait for the first snow or to visit the woods at night. Then they can know the invisible company they keep during the daylight hours.

Whales, painting by Jim Butler. Oil on canvas, 14 in. × 30 in.

Animals hold us to what is present: to who we are at the time, not who we've been or how our bank accounts describe us.

—Gretel Ehrlich, *The Solace of Open Spaces*

Animals are our kin, our fellow mobile inhabitants of the earth. From fish to apes, humans carry the characteristics of our ancestral heritage in our spines, hands, and the swing of our hips. Diverse body forms reveal the uniqueness of each individual species as well as shared traits. As we value the wildness of our relatives, such as the elegant liquid movements of the leopard, the fierce armor of the hermit crab and snapping turtle, the haunting voices of porpoises, can we see this diversity reflected in ourselves?

Humans, as animals, are part of the phylum of chordates and the subphylum of vertebrates, which includes all animals possessing an internal bony skeleton, most specifically a jointed backbone. Within the eighteen orders of living mammals, humans evolved from the order Primates, which includes tree shrews, lemurs, and anthropoids. Anthropoids further divide into New and Old World monkeys, of which we follow the Old World lineage. Humans and apes (hominoids) eventually separated from other Old World monkeys some 20 million years ago; humans (hominids) from apes some 5 million years ago. Our nearest ape cousins are the chimpanzees, with whom we share 99 percent of our genetic material.[1]

Like humans, primates rely on vision rather than on smell, which is dominant in most other mammals. They are adapted to tree living, with grasping

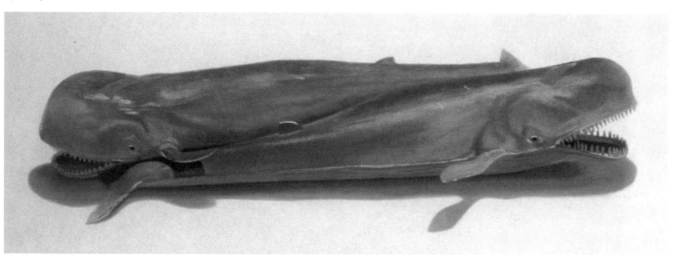

hands and feet for climbing. Primate mothers carry their babies on their bodies, developing a strong infant-mother bond. Primates eat insects, fruits, leaves, and occasionally lizards, tree frogs, birds, bird eggs, and smaller mammals. Humans retain all of the omnivorous habits of primates (eating everything), altered by cultural and economic forces that often determine what, how, and when we eat. Another important characteristic of primates is that we live in year-round social groups of both sexes and all ages. The ways primates use space and establish territorial boundaries are reflected in our locomotor system and social organizations.[2]

As we study an area with many animals, such as Wyoming grazing lands in the United States or the Masai Mara game reserve in Kenya, we first notice the large abundance of herbivores. In the American West, you can still see remnants of herds of migrating elk and the occasional reintroduced bison, plus smaller animals such as deer, groundhogs, rabbits, and mice, which provide food for foxes, coyotes, wolves, and an occasional grizzly bear. This species diversity reflects a balance of predator and prey, as well as sufficient migration corridors for stable and healthy animal populations. In the Masai Mara the presence of herbivores, such as the massive herds of migrating wildebeests and the all too infrequent rhinoceros, supports many carnivores, including mongooses, leopards, and cheetahs.

Animals live side by side. Observing hunting interactions in Kenya, one notices that the cheetah (the fastest of all animals) strides across the plains in full view of the zebras and gazelles. Prides of lions lounge under trees nearby, playing with offspring. The vulnerable prey remain alert but rarely bolt unless a cat is on the chase. The prey actually move with the predators, seeming to prefer keeping them in sight. The carnivores generally attack the elderly, injured, or feeble members of the herds during migration, ensuring the health of the species and, secondarily, the ecosystem. After a kill, many carnivores feed on a carcass, in a clearly defined hierarchy beginning with the primary hunter (sometimes replaced by a stronger carnivore), followed by scavengers such as the hyenas, and finally the vultures who pick the carcass clean until only the bones remain to slowly reabsorb into the earth.

This cyclical interdependence is reflected in the larger food web, which is composed of different trophic (nourishment) levels. Since photosynthetic plants, algae, and bacteria are producers that convert sunlight into energy to make their own food, they represent the first trophic level. The primary consumers are animals (herbivores) that feed on these plants, and the secondary consumers, carnivores, feed on the herbivores. Humans, as we have seen, are primarily omnivores (eating legumes and meat), although some, influenced by ethical, cultural, and economic motives, restrict diets to vegetarian fare. Scavengers feed on dead plant and animal bodies (examples are hyenas, turkey buzzards, and crows); and decomposers are fungi and bacteria that digest organic matter, essential to soil health (including microbes). Scavengers and decomposers play essential roles in cyclic life processes, cleaning the landscape and returning basic nutrients to the soil.

Keystone species are primary participants in an ecosystem. Their presence or absence determines the structure and balance of the whole system. A top

Otter, drawing by Laura Lee. Graphite.

The woodcocks are on their singing grounds. We hurry to arrive at the Robert Frost Trail, near Ripton, Vermont, by dusk. This is good habitat, a mix of alders and open field, to observe their display. As darkness falls, all is still; it is the end of March, perhaps it is too early in the season. Then comes the distinctive call, "beep, beep," followed by a high trill and dizzying freefall. We catch a glimpse of one mid-drop against the moon. Another passes so close to us, we jump. We are standing in their singing grounds. We back up quickly, exhilarated by proximity to this annual ritual, announcing life!

Caitlin Clarke

Drawing by Caitlin Clarke, age ten.

The bobcat crosses the road in front of us, as we anticipate an afternoon walk in Maine. We stop the truck as he runs quickly up a rocky slope and turns to watch us, calm and curious. I am most surprised by the tail—about five inches long, rather than the stump I had imagined from the name. Brownish fur and sharp eyes, a hint of spots. (A young male we later surmise, off to search for hares and small mammals, plentiful in this season.) My husband gets out and kneels in the gravel road-bed as they hold each other's gaze. I choose a side view, looking from the truck, stroking the texture of his fur with my eyes. When I lean forward, the cat thinks better of lingering, lopes off through a stand of black spruce, claiming pride of place.

Yoga: Cat Pose, Bidalasana.

predator, such as a wolf, can be a keystone species, maintaining limits on mammal populations. A beaver is a keystone species in a river system, shaping the structure and flow of waterways, influencing both habitat and food supply. As top predators, humans are an integral part of life processes. We are the primary threat to many animals, such as bobcats, wolves, and elephants. Since we have killed or removed most other top predators from the landscape, we now monitor populations of such animals as deer, caribou, and elk.

Because of our role as top predator, hunting is part of the ecological balance. Often, hunters know the woods and attend to the health of animals more than does the general public that buys Saran-wrapped meat in the supermarket. Although it is easy to denigrate hunters from the ease of our living rooms and kitchens, they are often the most knowledgeable spokespersons for animals and habitat. A novice hunter focuses on the animal; a skilled hunter includes the landscape in which the animal lives, climate shifts and food supply, reproductive patterns, and the health of the species in the bioregion to determine when, where, and how many animals to hunt. This large view is useful for all of us, acknowledging our roles in the larger process of life.[3]

Predator and prey interactions keep populations healthy and in balance with the land. This dynamic occurs naturally in landscapes if sufficient migration corridors are maintained and food supplies are intact. Bans on hunting, coupled with extinction of other predators, can cause havoc with the life cycles of various species. For example, overpopulations of deer strip forests, resulting in animal deaths due to insufficient food; overpopulations of snow geese threaten the tundra. Human interference with the balance of predator and prey requires a comprehensive view, recognizing the interconnections of flora and fauna and movement corridors in relation to sustainable populations.

Humans have complex relationships with animals, evolving through time, from being coinhabitants on a landscape to husbandry and dominance. Bison-like animals are documented in caves, showing magical invocation for the hunt. Wolves were perhaps the first animal to cross the line to pet; DNA studies show that wolves around human camps became partly domesticated as early as 100,000 years ago.[4] Cats and birds were featured in Egyptian and Chinese paintings and tombs. Spear tips and stone tools for scraping hides date hunting on the American continent to around 12,000 years ago, with possibilities of much earlier habitation. Profound change occurred in the relationship between humans and animals when hunting and gathering cultures changed to agrarian communities, focusing on domestication of both plants and animals within a particular locale, some 10,000 years ago. Most humans today consider that the function of animals on our planet is for human use. Domesticated animals, those raised and bred by humans for food, work, entertainment, or pets, now replace wild species (and the open grazing ranges of the past) as primary resources for food and for companionship.

Pets, in many cases, are a person's main contact with the natural world. Cats and other animals, birds, and fish in our homes, treated with awareness, can call forth affection and bring health to our bodies. Studies show that the presence of pets in nursing homes or hospitals helps in the process of healing. In many ways, animals remind us of our larger self. Yet domesticating animals often reduces their ability to survive in the wild without human intervention. Animals are bred for characteristics useful or entertaining to humans. This often undercuts survival instincts and fosters ignorance of the land and life processes in the wild. Many breeds of cows and sheep are literally too dumb to come in out of the rain or to birth their offspring without assistance.

Overbreeding and inbreeding dogs creates problems such as hip dysplasia. Cocker spaniels were bred for bulging eyes, because it looks "cute," but it is dangerous for the animal. Barnyard chickens and geese no longer fly. Domestication of animals instills an illusion of human control over the natural world, including a rather patronizing attitude toward all animals. Wild species are considered dangerous because they are a threat to domestication. For instance, we shoot coyotes because they will eat cats, young calves, and sheep; we shoot wolves because they will eat cattle (when starving); and we kill grizzly bears because they will kill us (when threatened).

As we value ease of control above intelligence and independence in animals, we are systematically removing wildness from our landscapes. Wildness reflects our own deep nature in a way that domesticated animals never can. Will we miss the powerful image of instinctual, alert independence and curiosity when the last Siberian tiger is shot (for his or her pelt) or the last white rhino poached (for the powdered aphrodisiac made from the horn)? Like the passenger pigeons that once filled our American skies and the gentle manatees now generally absent from our saltwater streams, many species are extinct or becoming so. Plants that provide food, habitat, and water for animals are being destroyed at an even more rapid rate, influencing the presence of wildness around us.

Animals' absence from a landscape warns of serious imbalance in the ecosystem. Silence, as Rachel Carson reminded us in *Silent Spring,* is a warning. Birds, particularly the sharp-sighted predators—eagles, osprey, and peregrine falcons—almost became extinct in the United States as pesticides (particularly DDT) were sprayed on fields, which made their eggshells so thin that offspring were born dead in the nests. Trout and salmon can live only in clean flowing water within a particular temperature range; a fishless stream warns us against swimming and drinking. Absence of songbirds identifies loss of habitat in far-reaching areas of migration. Soil without worms or beetles reflects overuse of herbicides and pesticides or erosion of nutrient-rich topsoil. Insects such as stonefly nymphs, mayfly nymphs, caddis fly larvae, dragonfly nymphs, and midge larvae (the favorites of fly fishers), plus worms, clams, and crayfish, indicate a healthy stream and river habitat. Although we should protect our wild animals solely for themselves and the health of our earth, humans too will suffer a great loss with their absence.

Nature writers such as Aldo Leopold and Barry Lopez have rehabilitated predators such as wolves, bears, and big cats in the public eye; Doug Peacock defends the lives and habitat of grizzlies; Mary Austin vivifies scavengers in her descriptions of buzzards; and Rupert Sheldrake reminds us of

It's bird hunting season again. Grouse are all over the logging roads on our daily walks. We heard them in early April on their drumming logs. Males beat wings in the air, propped back on their tails, claiming territory and calling a mate. A dry spring made for plentiful broods. Few babies died from heavy rains before pinfeathers formed. Now, after September storms, they come to the roads to dust themselves. With this quantity of birds, some won't survive the winter; there's not enough food in the overcut forest.

So we will have grouse, or partridge as they say, for Thanksgiving and Christmas dinner. Raised on a farm, where the pig became our bacon, and the hens, fed in the morning, became dinner, I understand. Richard Nelson, in Heart and Blood: Living with Deer in America, *writes: "I've come here to hunt only with my eyes, a luxury of the twentieth-century world, where freezers and grocery stores foster the illusion that life sustains itself without taking another life."*

Our backyard is a training ground, with Birmingham Roller pigeons, two Labs, and our Brittany. My husband and neighbors are trying to teach the dogs not to jump the birds, to flush or point but not pounce. When the pigeons are released from their cages, they do a few spins and return to their loft unharmed. Hunting dogs are a body behind a nose, the training book notes, using scent for detection. They have 125 million to 1 billion smell cells to our 5 million! But grouse, well disguised in fall foliage, with zigzagging flight patterns through the trees when disturbed, are generally safe. Bird hunting, in our family, mostly means long walks in the woods with the dog.

Abenaki scholar and storyteller Joseph Bruchac said about hunting: "If you are going to learn how to hunt, you'd better know how to pray."

the instinctual prowess of dogs and all they can teach us about the larger patterns of life on earth. To read their encounters with animals is to feel connected once again to who we are and where we came from. Animals are our kin. In an era of rapid change and alteration of habitat, our awareness of animals is essential: those at all levels of the food chain, as components of the human psyche and as reflections of our own instinctive nature. They are our daily companions, be it the dog at our side or the wild elk on the distant horizon. Our attitudes affect the lives and deaths and futures of our animal relatives.

⚮   ⚮   ⚮

TO DO

Lateral line*

10 minutes

Lying on the floor, eyes closed:
• Stroke your hand down the sides of your body to stimulate sensation. Pay particular attention to the armpit, which has many lymph nodes.
• Imagine you have a fish body, a head and tail (with no arms and legs) connected by a flexible, cartilaginous stiffening rod called the primitive notochord.
• Roll onto your side; find a comfortable way to stimulate your side body. This is called the lateral line in fish, a particularly sensitive area allowing fish to swim in schools.
• Stretch along one side to roll onto your back, then to the other side, and repeat.
• Lying on your belly: focus your imagination on the cartilaginous stiffening rod—the primitive notochord. Curve your head toward the tail on one side (a C shape) and feel lateral flexion; try the other side. (You are using the multifidus muscles on the back and sides of the spine, which allow subtlety of spinal movement.)
• Wiggle your tail! Feel the undulations ripple up your spine.
• Continue to feel the mobility of your spine as you transition to stand in postural alignment. Add lateral awareness, extending out through the sides of your body. Stretch your arms to the side, feeling this connection.

Extending your lateral line connects you to the side so that you can feel others in community, lateralizing your perception. Notice this when you are sitting next to someone or in a circle with a group.

*Title and exercise drawn from Caryn McHose, movement teacher and educational bodyworker.

Reading the signs: Ten questions to ask an animal

2–4 hours

Using field guides and place notes, write about an animal at your place:

1. Where do you live, and what kind of home do you make? (Consider habitat, distribution, social interactions, and materials.)

2. What is your relationship to the seasons? (Consider patterns of migration, hibernation, temperature control, water, and plants.)

3. What do you eat, and how do you find or catch it? (Reflect on the food chain, teeth, movement characteristics, and senses.)

4. Who are your predators, and how do you protect yourself? (Same as above.)

5. What are your mating characteristics? (Consider behavior, timing of cycles, relationship to mate, offspring, and care of the young.)

6. Are you a social animal, or do you live alone?

7. How old is your species; who are your relatives, and how long have you lived here?

8. Do you have a sound or language that others hear?

9. How have humans affected your life, and what is a day like for you?

10. What about aesthetics? (Consider sound, color, shape, textures, movement, scent, and habitat.)

Photograph by Erik Borg.

Place visit: Observe an animal

Visit your place at dusk or dawn, preferably after a fresh snow or rain so that footprints will be visible. Approach as silently as you can, and notice animal signs at your place. Observe one animal for twenty minutes. Follow tracks, look for holes or nests and scat. Consider the relationships. If you see nothing, wait until your senses have adapted to the place and the animals feel comfortable with your presence. Remember, most animals smell and hear better than you do and have significant reason to fear humans. As you look for them, they are watching you. Birds, squirrels, and your own dog can be fun to watch. Imagine doing this each day for a week, a year, a decade, and a lifetime. 20 min. Write a story as though you are experiencing the world from inside that animal. 10 min. Read aloud to yourself or a group.

CROSSING OVER: FLOATING

A red-tailed hawk floats overhead as we paddle a northern stretch of the Connecticut River in fall. A week ago my husband and stepson flew in a glider, following the updraft of warm air under cumulus clouds along with these migrating birds. Today we gaze up at the clouds and relish silence. Stopping for lunch on a sandbar, we listen to the song of crickets, a distant dog, water over riffles.

I feel a shadow on my skin and see a bird swoop through the air, right over my belly. Detecting flecks of brown, white cheeks, and talons on feet, I realize that it is not a hawk, as I first imagined, but an osprey with a five-foot wing span. Hearing "tu tu tu" as its call, my husband is encouraged to pick up his rod and fish alongside this sharp-eyed expert.

Our Brittany, Tobie, puts his paws on my notebook; he wants to be off exploring—his nostrils and air pouches flare, filling with scent. All these senses— the coolness of air on skin, patterns of shadow and light, sounds and vibrations, unimagined scent—merge into my awareness of the moment, telling me of all the life that inhabits this one private place on the 407-mile-long Connecticut River.

Before the first frost, before the hurricane winds coming up from the panhandle that will pour rain, wind, and even sleet on our house this very night, before we enter the next phase of my father's illness, which will end this fall, there is this time of peace, this time with the wings of the osprey, telling of ease.

Read aloud, or write and speak your own story about animals.

⧉ DAY 20

Endocrine System

Perhaps what we fear most is nature, even our own.

—Terry Tempest Williams, *Leap*

The endocrine system can be considered our system of transformation. Mental and physical states can change almost instantaneously or over hours, weeks, or years through secretions of chemicals called hormones into the bloodstream. The endocrine system works as a complimentary system to the nervous system to integrate, focus, and harmonize the body. Chemical in nature, the endocrine system existed before the nervous system and acts in a diffuse way throughout the body, affecting a wide range of processes,

Leeds Dam. Photograph by Bill Arnold.

In a Body-Mind Centering workshop on the endocrine glands and yoga, a teacher put her palms on the front and back of my ribs. Locating the pancreas halfway up between the belly button and sternum, she said encouragingly, "Move from comfort first; expression will follow." When she removed her hands, I began exploring on my own, stretching here, twisting there. Following the flow of internal movement, a sweet song emerged. I walked around the room for half an hour, singing. Remembering the many years in choir, struggling to produce sound, I was amused. All along, the pancreas had a voice of its own!

Learning to do a handstand in yoga, I fear that my arms will collapse. When I study the endocrine system, I add energy from my tail (coccyx), extending down through my feet; energy from my heart extending out through the hands; energy from the pancreas connecting hands and feet to my center; extension from my skull and eyes linking me to place. Up I go.

I taught dance to Anya and Nell when they were children: tendues and pliés holding the backs of folding chairs. When they were thirteen, I offered to choreograph a solo with each. Nell and I created an interior world, private. Dips and dives, spins and circles signaled life lived in the moment. Without hesitation, she created the whole, beginning to end. For Anya, I brought two heavy granite rocks, perfectly round, and placed one in each hand. She hit them together, and the dance became a ritual; each time the structured work is performed, it has a life of its own. Both of these possibilities are part of a dancing life, creativity lived in the moment, and spontaneity unleashed within repeated form. This ceremony created a container for the powerful energies of their adolescent bodies, energies that will accompany them throughout their lives.

whereas the nervous system acts within milliseconds, usually upon a very specific target (e.g. a muscle), and generally in the short term rather than the long term. Through its connection with fluids, the endocrine system blends the body into an interconnected whole, modulating reproduction, metabolism, and responsiveness to the environment.

Endocrine glands and organs are located in various single and paired structures along the front surface of the spine. Beginning with the pelvis and moving up toward the head, the following structures are currently recognized as belonging to the endocrine system because they release known hormones into the bloodstream: *reproductive glands* (two ovaries, testes); *adrenal glands* (two glands, located atop the paired kidneys); *pancreas* (below the breastbone, about six inches in length, extending from the front to the back of the body); *heart* (behind the sternum, between the lungs, reclassified in 1983 as an endocrine gland)[1]; *thymus gland* (above the heart); *thyroid gland* (two lobes, located in the neck); *parathyroid glands* (two glands, located on the posterior surface of the thyroid); *pituitary gland* (two lobes, suspended by a stalk from the hypothalamus in the brain); and *pineal gland* (single structure, attached to the thalamus of the brain). The kidney, digestive tract, and placenta have known endocrine functions as well. Every cell in the body has the capacity to respond to endocrine messages.

Bonnie Bainbridge Cohen, in *Sensing, Feeling, and Action*, also discusses the *coccygeal body* (a small cluster of cells at the interior tip of the coccyx), the *carotid bodies* (on each side of the neck along the carotid artery), and the *mammillary bodies*, in the center of the brain. At an experiential level these tissue bodies have characteristics of the endocrine system; they also are highly vascularized, which suggests that they are likely to secrete hormones into the blood. Explore all the glands for yourself and notice what you find. Together, these organs and tissues of the endocrine system correspond roughly to the seven major chakras in yoga.[2]

Hormones are chemicals secreted by one cell that affect the functioning of another cell. Hormones exert their influence by altering metabolic processes (rate of protein or enzyme synthesis, enzyme activities, or molecular transport through cell membranes). Their effect is restricted to specific target cells—those cells with specific receptor sites for the hormone molecule. Hormones are very potent, affecting changes in target cells even with low concentrations. Thus, as a group, endocrine glands and their hormones control the rates of chemical reactions, aid in transport of substances through membranes, and regulate water and electrolyte balance. They also play vital roles in reproductive processes, development and growth, and regulation of body rhythms. For example, the pineal gland secretes melatonin in response to decrease in light registered in the retina of the eyes, governing our circadian rhythms in relation to cycles of day and night. In many mammals this includes seasonal fertility and infertility cycles.[3]

Endocrine support, described in terms of energetic frequencies, has the highest frequency in the body. Particularly when adrenaline-like hormones are released in balance with other hormones, there is a sense of vitality, a buzz, a lightness unique to the endocrine system—beyond neuromuscular "effort." During endurance hiking or kayaking, this is sometimes described as being "in the flow." These hormones enhance activity by increasing metabolism and have a great variety of effects, such as on blood pressure and heart rate. You can see endocrine activity visually in athletes or dancers—the elevated jump of Michael Jordan in basketball, the seemingly effortless display of a perfect gymnastic routine, and the transcendent leap by Baryshnikov in ballet reflect the pervasive unity of endocrine support. Try jumping with your muscles; then activate the endocrine system and notice the difference in lightness and energetic output. Interruption of this energy flow, linking tail with head through all the connecting structures and tissues, results in blockage of expression. This can be experienced as sluggishness, lethargy, or scattered, unfocused energy.

Endocrine tissues, generally tiny communities of secretory cells surrounded by blood vessels, have tremendous influence on the body. Because of their force, it is important to balance their activity in our bodies. With attention, it is relatively easy to distinguish the qualities of endocrine activity through sensations. From an experiential standpoint, you can explore the physical function of each endocrine gland and its accompanying feeling state to determine the vitality of this tissue in your own structure. Similarly, if there is an area of the body that feels blocked, it is possible to stimulate the endocrine glands through sound, visualization techniques, or movement to increase circulation and vitality.

The hypothalamus of the brain is the keystone for the endocrine system and regulates the autonomic nervous system (fight, flight, freeze, and friendly response). In balanced situations, the endocrines provide alertness and engagement with the environment. In situations perceived as stressful, however, the hypothalamus sends a chemical message to the adrenal glands to release cortisol to prepare the body for physical activity at the level of survival. Some stress can be useful: it sharpens the responses and makes one more alert. Generally, though, long-term, continual stress can be accompanied by a decrease in the number of white blood cells in the body, which reduces resistance to infectious diseases and to the growth of some cancers. Prolonged stress also can cause high blood pressure and ulcers. Thus, our perception of a situation plays a key role in endocrine function. This is one reason that meditation has proved effective in reducing cortisol levels and high blood pressure; that walking, physical activity, and yoga are useful for reducing depression.[4]

Energy and energetic states in the body are hard to describe in traditional Western scientific terms, although their presence and effects in the body are easily measurable. Cortisol levels, for example, can be monitored through tests on saliva. Because every long-lasting movement form is based on anatomical truth, traditional movement forms have evolved highly specific language to describe the causes and sensations of energy in the body. The traditional Indian form of yoga, two thousand to five thousand years old, has elucidated the energetic centers along the front of the spine as the chakra system, with postures designed to stimulate each gland; the energy or life force is called *prana*. Chinese acupuncture and acupressure work with energy

I learned about the endocrine system through performance. On stage, adrenaline moves through and animates the moment. When I am dancing well, I become a vehicle for audience members to experience their own bodies moving beyond the norm. Various life events bring these same endocrine glands to the forefront—a car accident, for example, sexual climax, menopause, a fight. In a moment of clarity, what we do with those coursing energies is unique to the moment. Forms have been created throughout history to channel energy and investigate this vast resource for health and transformation rather than destruction and dissipation. In essence, it is about consciousness. We wake up, or we don't.

Tai Ji: Move Hands Like Clouds

meridians, and each organ is associated with a pulse. Specific reflex points help move energy through the body, harmonizing the flow; the energy or life force is called qi (chi).[5]

In techniques that probe the depths of the body, energy is considered paramount, its interactions with mind to be carefully monitored. From shamanic practices to Qi Gong and Zen meditation, different focusing methods describe the effects of energy and how to channel it toward health and ethical action. Lifelong involvement in contemplative practices combines experience of specific energetic states with expression through disciplined form. The techniques introduced in this text, such as authentic movement, yoga, t'ai chi, meditation, and contact improvisation, can be studied to enhance awareness of the dialogue. We can refer to these disciplined techniques to notice and experience the structures involved. In our daily lives, a balance of familiar activities, such as lively walking and rest, singing and silence, help to move the blood and balance the glands.

⊁ ⊁ ⊁

TO DO

Balancing the endocrine system with sound
20 minutes

Standing, eyes open or closed:
• Send a high-pitched tone (use a vowel sound like *ah, oh, ee*) into the very top of your skull. You may want to jump as you sound to help vibrate into your head (straight up, like a Masai warrior).
• Roll down the spine and include a falling sound as you move, gradually decreasing in pitch as you articulate lower in your body. Focus the vibration of sound in each moving part of your body.
• When you are upside down, find the lowest tones you can make. Stomp your feet a few times to stimulate the lower glands in your pelvis.
• Swing your pelvis so that the spine and skull swing like a clapper in a bell; release your jaw and neck.
• Reverse, going from low to high tones as you roll up. Finish by jumping as you make the highest tones. Throw your arms up into the air for added momentum. Notice emotional connections! Pause in open attention to feel the sensations stimulated in your body.
• Repeat several times, observing places where sound has difficulty registering in your structure or where rolling down the spine is restricted. Pause in these areas and stimulate them with sound.

Sound and movement vibrate and stimulate the glands in the body, opening areas of compression and restriction. As you equalize attention to all parts of the body, you help balance the glands.[6]

Charlotte Sikes, dancer. Photograph by Sommerville B. Johnston.

Yoga: Modified Bridge Pose, Setu Bandhasana.

Elevated pelvis (Modified Bridge Pose)
10 min.

Lying in constructive rest, eyes closed:
• Lift your pelvis off the floor so that it creates a long diagonal from your knees to your shoulders. Keep the neck free, chin slightly toward the chest, as though a river is running down your core and through your neck and head.
• Breathe deeply, allowing the organs to relax toward your chest.
• Lower the spine slowly, vertebra by vertebra, elongating your pelvis toward your heels. (Be sure the knees remain parallel to each other.)
• Repeat, lifting the hips slightly higher and allowing weight to rest on the top of your shoulder girdle, without compressing the neck.
• Imagine the endocrine glands and tissues along the front of your spine, tail to top of the head. The thyroid gland in the neck is particularly stimulated by this pose.
• As you lower the spine, release each gland toward gravity: thymus (base of neck), heart (center of ribs), pancreas (below ribs), adrenals (back of floating ribs), gonads (ovaries in women, testes in men), and the base of your coccyx.
• Lying in constructive rest, allow the back body to support the organs and glands.
  In yoga, this inversion is called Setu Bandhasana, Bridge Pose. Visualize the span of your spine, linking head to tail.

Place visit: Attention to the endocrine system

Sitting, with eyes closed, wriggle your hips or move the buttock flesh slightly away from your sit bones, so that you can feel the ground. Scan through the body, balancing front body and back body muscle tone. Then scan through the core of your body, imagining a light moving up the front of your spine. Begin with your pelvis and continue through each area in the torso, making space in the body with breath as you notice the endocrine system. Then chant *AUM,* allowing the low tones (A) to vibrate in your belly, the middle tones (U) to move through your chest, and the (M) to tingle your lips and skull. Does vibrating the endocrine system change your awareness of place? 20 min. Write about your experience. 10 min.

Insects

The sound of katydids in the evening light . . .
— Vincent Dethier, *Crickets and Katydids, Concerts and Solos*

Insects undergo complete or simple metamorphosis, changing body form one or more times from egg to adult. They are among the most populous animals on earth, comprising over half of all living things on the planet. In fact, one in every four species of animals is a beetle. Like the endocrine system, insects are the "small" affecting the whole. Both are often overlooked, denigrated, or out of balance in our fast-paced world, yet they are essential. Many plants and animals depend on insects directly or indirectly for food. The diversity of insects is directly proportional to the diversity of life on earth; insects pollinate approximately 80 percent of flowering plants and facilitate recycling of organic matter and soil aeration.

Insects reveal extraordinary variation in color, form, and size. We can be delighted by their beauty (eastern tiger swallowtail butterfly and green darner dragonfly), surprised by their shapes (praying mantis and northern walkingstick), bewildered by their size (no-see-ums and snow fleas), lulled by their sounds (true katydid—katy-*did* and katy-*didn't* calls at night—and field crickets' triple chirps), and confounded by their quick antics (whirligig beetles on the surface of ponds and hummingbird moths around plants). They are vital aspects of the world we inhabit, occur worldwide (very few are in oceans), and indicate the ecological health of other systems.[1]

Insects are classified in the kingdom Animalia and the phylum Arthropoda, identified by an outer waxy skeleton and jointed legs. Of the various classes in this phylum, several are often mistakenly grouped together in colloquial speech under the term insect or bug. *Crustaceans* include terrestrial pillbugs and sowbugs found under rocks and rotting logs, as well as many marine and freshwater forms. *Millipedes and centipedes* have many legs—two pairs per segment for plant-eating millipedes and one pair per segment for predatory centipedes. *Arachnids* include eight-legged spiders, daddy longlegs, ticks, mites, and scorpions.[2] *Insects,* the topic of this chapter, include six-legged dragonflies, grasshoppers, beetles, flies, butterflies, ants, wasps, and bees as cohabitants on the planet. Although the total number of insect species is unknown, estimates suggest that at least 600,000 have been identified as belonging to one of thirty-one taxonomic orders, with over 88,600 species in North America alone. Only one of those orders, Heteroptera (also called Hemiptera), comprises the true "bugs," with sharp beaks, such as the kissing bugs in South America or the giant water beetles that are

Chest of drawers, by Kristina Madsen. Pearwood, gesso. Photograph by David Stansbury.

"Mariposa, Mariposa!" a two-year-old boy named Xun calls out. He is pointing at a monarch butterfly emerging from its chrysalis. Visiting from Chiapas, Mexico, his parents were once students at this college. Now we sit together with my class, surrounding a glass aquarium. This has been the home for some twenty caterpillars until a few days ago. Now pale green cocoons (chrysalides) trimmed in details of gold, hang from the screen grid used as a lid. Remnants of second-growth milkweed lie in the bottom, their favorite food, abandoned. One by one, day by day, the chrysalides darken and break; butterflies slip out of their casings, unfold their wings, and flutter. Today, we carry one outside, into the sunny autumn air, and place it on a maple leaf. "Mariposa," Xun repeats waving good-bye. Butterfly! From Middlebury, Vermont, this orange and black wonder will gather with others to fly to mountains in central Mexico for winter, a distance of thousands of miles. Both Vermont and Mexico are home, for Xun and for the Mariposa.

a delicacy in Thailand; bugs found in North America include water striders, stink bugs, leaf bugs, bedbugs, and many others.[3]

Essential components of the food web for fishes, birds, and other wildlife, insects have both harmful and beneficial effects on humans. Less than 1 percent of insects are harmful. Even though this group is small, insect "pests" can cause losses averaging from 5 percent to 15 percent of the annual agricultural production, and they challenge the endurance of even the most resolute organic farmer or gardener. To witness your wheat field devoured by aphids, your potato patch eaten by potato bugs, or roses and raspberries decimated by Japanese beetles creates more than a moment of pause. Objectively, however, the income loss to farmers through insects is estimated to be overshadowed by the cost benefits associated with pollination. Some mosquitoes and ticks also pose health threats by aiding the spread of disease; termites threaten homes by eating and digesting wood.[4]

In the positive view, insects can be useful, resourceful, and aesthetically beautiful. They produce many beneficial products: the silkworm spins silk inside its cocoon; and bees drink nectar of flowers, remove the water, and regurgitate it to produce honey, stored in combs made of useful beeswax. The pollinating activity of wasps and bees (with taste and smell receptors on their feet and legs) helps in the production of fruits, vegetables, flowers, and seeds. Beetles (and their larvae, grubs), ants, and grasshoppers provide a food resource in some cultures. Certain insects eat other "harmful" insects: the larvae of ladybugs eat aphids, some beetles eat gypsy moth larvae and cutworms. They assist in the essential decomposition of dead plants and animals; a group of dermestid beetles in Costa Rica can strip a carcass to the bone in several hours; a red and black burying beetle in Maryland can carry a dead mouse on its back to a sandy spot, then bury it and lay eggs in its carcass, providing food for its young. Insects are thus an essential component of topsoil; ants in particular help redistribute and aerate the soil. Many insects are beautiful: observe a four-inch pale green luna moth, with elegant tail, resting in a beech tree with its white-etched wings outstretched. Waiting to fly at night, it offers an elegant display of form and color.

One evolutionary characteristic of insects is their relatively small size. Prehistoric dragonflies (320 million years ago) featured wing spans forty inches wide, like small pterodactyls. As life evolved, insects found their survival niche: most range from less than a millimeter long to around six inches, with over half smaller than one-quarter inch in length. Their small size allows them to thrive on food that is overlooked by most vertebrates. Some beetles can eat such unlikely foods as opium, strychnine, and deadly nightshade! Insects also live in places unreachable by larger animals and inhabit every ecosystem except the sea. Whereas the human species has lived in its current form for approximately seventy-five thousand years, certain insects, such as cockroaches, have maintained the same successful shape for the past three hundred million years; mosquitoes, for nearly a million years.[5]

Another characteristic of insects is their capacity for transformation from egg to adult, known as metamorphosis. *Complete metamorphosis*, unique to insects, begins with eggs (laid by adults), from which larvae are hatched.

Larvae feed and grow, sloughing their skin several times, until they find hidden places to camouflage their immobile pupal state. In this pupal phase, larval organs mysteriously dissolve, adult organs develop, and the adult eventually emerges. For example, if a monarch butterfly begins its life in New England as an egg, then it becomes a rotund, yellow, white, and black–striped caterpillar (often found on milkweed plants) and attaches itself to the bottom of a convenient leaf (like beet greens in your garden). There it transforms into a pale green, gold-specked chrysalis and finally emerges to become an elegant orange-and-black–veined butterfly, which will migrate south from North America to overwinter in the mountains of central Mexico. Its offspring, one or two generations later, will arrive back at the northern site, and one or two generations will pass before the next migration repeats.

Simple metamorphosis in insects such as dragonflies, grasshoppers, and bugs skips the pupal stage, so the insect gradually develops from hatched nymph into adult. Thus, simple metamorphosis is egg, nymph, adult. The nymphs usually resemble the adults, like the grasshopper, except in size and development of wings. Another more extreme example is the hellgrammite nymph, an aquatic predator; generally found under rocks in streams, it is a favorite food of bass. The hellgrammite nymph is the larval stage of the two-inch-long eastern dobsonfly adult, with sharp mandibles (in the male) three times as long as its head. It is generally found flying over nearby woods or attracted to lights.[6]

Every insect is covered with an exoskeleton, a thin, waxy, acid-resistant outer skeleton that protects the insect from dryness, humidity, and disease organisms. Insects have three main body parts: the head, the thorax, and the abdomen, with other organs attached. The segments are composed of several hardened plates called sclerites, connected by soft, flexible intersegmental membranes. The head serves as a sensory center composed of mouth parts (for sucking, chewing, biting, piercing, and/or licking), eyes (generally two kinds: compound—sensitive to movement—and simple—sensitive to light and dark), and antennae of various lengths (sometimes longer than the body). There are six legs on the thorax, generally four wings, and one pair of antennae, sensitive to vibration and balance, on the head. The abdomen is usually the largest segment, housing internal organs, including the reproductive system.

Many variations on form and function exist. Names for insect orders often end in *ptera* (Greek, *pteron*, meaning "wing"), and wing characteristics assist in identification. True flies have only one pair of wings; beetles, such as ladybugs, have a wing turned into armor covering a second wing underneath. Wing movement in insects is produced primarily by changing the shape of the thorax, because wing muscles attach to the thorax walls rather than to the wing base.[7] Insects also can dry out easily; they have an open circulatory system and open holes on their sides for respiration; the latter let in gases (oxygen goes directly from the outside to tissues through tracheal tubes, rather than through the blood, as in vertebrates), but these also dry the insect out. Thus, they frequently live under rocks, near water, and in leaf-litter, emerging in the cool dampness of evening.

Shadow darners, the "big ones" of the dragonflies, are flying up and down the logging roads, feeding on insects. I watch one catch a moth in midflight and spit it out. A meadow hawk dragonfly lands on my knee, one of the last of the year. Katydids and field crickets make an endless chorus. Red milkweed beetles ride on our clothes. Wasps gather on the south-facing side of our house; ladybugs move indoors. It is September in Maine. Winter is coming.

I love honey. Friends bring me jars from their travels. Today, for morning tea, I chose from five containers on my shelf: the largest and most exotic, carried by ten-year-old Sam all the way from Samos, Greece, plus several lighter Maine varieties: Mrs. Hall's from Mt. Vernon, Swan's Wild Raspberry from Brewer, Abnaki Apiaries providing comb honey from Wellington, and my current mainstay, identified simply by the handwritten label: Bees Honey, from the Smiths in Aroostook county. Each has its own subtle color and flavor determined by soil, plants, and time of collection. So when a honey bee lolls in the orange depths of the squash blossom or buzzes in the apple blossoms or tastes the tiny stamens on the elegant poppy with its feet, I try not to disturb these little creatures hard at work ensuring the cross-fertilization of plants and also producing my favorite food.

Entomology, the study of insects, has increased our understanding of many other subjects, such as evolution and genetics, ecology, and sociobiology. Insects are, however, hard to get to know. As we have seen, there are difficulties with scale, including life span and size. They have complex life cycles; it is a challenge to recognize the cutworm, eating your broccoli plant, in all its four stages of development: as egg, caterpillar, pupa, and adult moth. Some insects overwinter in each stage; in northern climates, many will not complete their development unless they have a period of dormancy at cold temperatures. Their life spans vary: some insects regularly have two generations a year, while others have several, and a few require more than a year to develop. Most insect adults live only a few days or weeks, although some adults overwinter and certain social insect queens (for instance, Honey Bees) survive for years.[8] The seventeen-year locust is the most extravagant, spending seventeen years as a nymph hidden in the ground (sucking sap from a tree's roots) before climbing the trunk and buzzing in the tree for a week as an adult.

The sheer numbers and diversity of insects can be overwhelming. Insects outnumber other higher animals in surprising proportions; you can find as many as several million per acre of soil, occurring at weights of up to two thousand pounds. (In contrast, birds add up to about one pound per acre.) In fact, there are five thousand to ten thousand different kinds of caterpillars in North America alone, each with its own unique characteristics. Insects have highly adept perceptual and motor systems; they can hear ultrasound, see ultraviolet, and fly up to thirty miles per hour. Insects also respond to almost imperceptible changes in the environment; watch for the increased movement of ants to foretell rain, bees to predict the first frost. For flyfishers who tie their own flies, the study of water insects in various phases of maturation leads to the ability to replicate insects and thus attract specific fish (often as catch-and-release). Beekeepers, too, have a special relationship to insects, studying their habits and cycles so that both bees and humans remain uninjured in the process of gathering honey.

Insects can also inspire us with their work. They are stonemasons (primitive caddisflies in streams make turtle-shaped cases of small pebbles that they carry around with them), diggers (dung beetles and tumblebugs work in pairs to form a mass of dung into a ball, dig a hole, bury and lay eggs in it, providing food for their young), weavers (black fly larvae make a vase-shaped cocoon on rocks, weaving thread extruded from a cone near their head; when scared, they can lower the line into the water, and go away from their rock, then pull themselves back), carpenters (black carpenter ants drill, cut, and make galleries), sappers (all the aphids, tree hoppers, leaf hoppers), makers of paper (northern paper wasp and bald-faced hornets create elaborate paper homes suspended from branches or porch eaves by eating and digesting wood) and producers of silk (cecropia moth of the giant silkworm moth family). Although insects are diverse and their habitat far-reaching, all of these insects can be found during a summer in New England, and in many other regions of the earth.[9]

Insect control is big business in the United States, resulting in the manufacture of various chemicals that make life difficult for insects and other animals, including humans. Regardless of the fact that we have tried mightily to kill insect populations, we are rarely successful in exterminating insect pests. Organochlorines are a group of insecticides that originated out of wartime experiments. They are dangerous because of their persistence in the environment and their severe impact on wildlife. (DDT, a popular organochlorine from the 1940s through the 1960s, was banned from the United States in 1972 but is still made in the United States for export.) Organochlorines are broad-spectrum insecticides, wiping out useful insects along with the harmful ones, the predators as well as the prey. Spraying DDT kills the birds (by thinning the eggshells) that eat the insects, just as it kills the dragonflies that eat the mosquitoes. Concerned individuals are attempting to break the chemical cycles introduced at midcentury by returning to diversified crops, supporting organic farms, and encouraging the public to accept some blemishes on fruits and vegetables (caused by a scale insect affecting the surface but not the fruit inside).[10]

Perhaps insects are irritating to humans because they elude our control, thwart categorization, and slip by our perceptual gateways. They also threaten our sense of dominance by sheer number, volume, and staying power. But as we shift to a relational perspective, we can remember that most insects are not harmful, and they play essential roles in the health of the ecosystem. Each life phase of a moth, for example, offers food to other animals (birds, frogs, fish, even humans). It seems more useful to acknowledge the variety, beauty, and adaptability of insects—characteristics worth emulating!

⚹ ⚹ ⚹

Noticing insects

20 minutes

• Take a walk, bringing your attention to insects: Where do you find them? What do they look like? Be as detailed as possible. Make drawings to remember species so that you can look them up and learn their names. If you have a magnifying glass, bring it along to study the complex and beautiful forms.

• In fields, look in tall grasses or near the soil. Bring a glass jar (or better yet, a butterfly net). Swish it through grasses to gather a collection. Observe live insects for a few minutes, then let them go.

• In the woods, look under leaf litter or in the duff; check under bark, on vegetation, in dead branches. Dig in the earth; notice who's there and in what stage of development.

• Visit a stream or pond: Notice the insect life present. Pick up a rock and observe what you find. Look at understream boulders, trying not to disturb the habitat. Reflect on the food chain, the various life cycles of insects.

Remember, insects aerate the soil, break down dead plants and animals, and have various forms in different seasons. Most are not harmful to humans.

The spider crawls up beside me on my lakeside rock in Maine. I barely notice as it maneuvers eight legs across the landscape of my skin. Thinking a pesky fly has come to visit, I jerk my arm, then continue dreaming. A few minutes later, a tickle, another flick, a moment's irritation. The third time I kick my foot and sit up, peering around me to know my inquisitor. A big, furry spider swims determinedly toward my rock, and I jump to my feet, relinquishing my ground. The false eyes on its back, a deterrent to birds, also keep me at bay, as does its defiant forward legs signaling distance. This rock was inhabited before my arrival, and Woolfie, the wolf spider, claims pride of place. I've shared beaches and homes with spiders all summer it seems, their intricate webs holding me to this place. House spiders appear on curtains and stairways, and barn spiders elaborately decorate the ceiling of our porch. These arachnids devour many insects, such as flies and mosquitoes, and are self-initiated proprietors of a tidy house. I've had to rethink my place in their world, and theirs in mine.

TO DO

Monarch caterpillar. Photograph by Erik Borg.

CROSSING OVER: FIRST VISITS—1

When I ask him to show me his room, he takes me to a grove of trees. Returning indoors, he refuses to recognize his bed. The blankets are on the floor; no one can tell me what happened. They hand me an envelope holding a small pocketknife, fingernail clippers, and a dollar bill. I am scolded for letting him bring these objects to the Alzheimer's home, and I take them back quietly, thinking, Dad knows how to hide things.

As we walk down the hall arm in arm, I ask him to name the painters whose prints are on display. "Like we always did in museums," I prod. He says, "Honey, I don't know things like that anymore. I barely know who I am." When we return to his room, he gestures at his own paintings of birch trees and boats, recently hung. "These are the good ones," he says smiling.

"Sweede. Sweede," the aides call after him altering his nickname. "Sweede," like "sweet." Three giggling island women point at us as we walk, "You look so much alike." They gesture and wave fingers in the air to link us. I see my gray hair, wrinkles, the stoop of my spine hovering next to my father. I imagine a cocoon surrounding and protecting us, as he begins withdrawing from this life and opening to the next.

Read aloud or write and speak your own story about insects.

Insect stories
20 minutes

Write a short anecdote about insects. An anecdote is a story with a beginning, middle, and end. It may be a story you have told before, a childhood memory or a recent event. 5 min.

Write about someone who taught you about insects. 5 min.

Create a story as though you are the insect. 5 min.

Read aloud to someone else. 5 min.

Place visit: Attention to insects

Seated or moving around, in open attention: Learn the names and characteristics of two insects at your place. Choose one and experience your place from the perspective of this insect. Where in this place would you spend your time; what part of your life cycle would you be in (what form) during this season; and how would you perceive the world? What would you eat and what would you do with the human visitor? 20 min. Write about your experience. 10 min.

DAY 22

Digestion and Nutrition

We sacrifice so much for flatness.

—Caryn McHose, evolutionary movement educator

Digestion incorporates the nutrients of the earth into the tissues of our bodies. Occurring successively in each digestive organ, digestion is a process. Most of digestion takes place in the small intestines over a period of three to ten hours. Several muscular rings, or sphincters, link distinct parts of the digestive tract. There is a choice at each sphincter to go forward or to hold on, reflecting larger implications of taking in, integrating, and letting go. Our attitudes about food affect all aspects of the experience. When the nourishment cycle is completed, digestive processes return nutrients back to the earth.

The digestive system is one continuous tract from mouth to anus. Reflecting evolutionary origins, this passageway allows elements from the outer environment to pass through the inner environment of the body for absorption and elimination. The process begins with the *nose and mouth*, including *lips, teeth, and tongue.* The chemical receptors in the nose for smell (the oldest of the special senses) are closely linked with the taste buds in the tongue, which register bitter, sour, salt, and sweet flavors, and with emotional memory, influencing our receptivity to food. The teeth begin mastication of the food in conjunction with secretions of the salivary glands and movement of the tongue. The *throat*, or pharynx, is the first sphincter, simultaneously opening to the esophagus and closing the adjacent passageways to the trachea and nose. The *esophagus*, a long tube leading to the stomach, fills the space between the front of the cervical vertebrae and the back of the trachea,

Can Opener, painting by Jim Butler. Oil on canvas, 30 in. × 120 in.

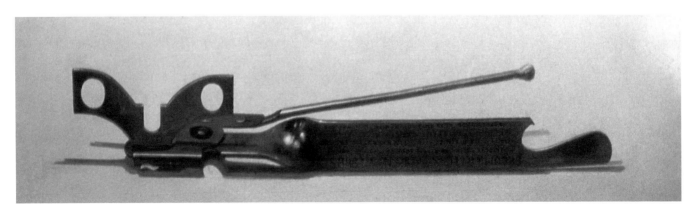

FOOD

"Food: The Ultimate Killer!" This heading of a student paper woke me from the doldrums of grading. Photos in the appendix depicted various pencil-thin models: the familiar anorexic ideal perpetrated by our culture, the pressure to be thin at any cost to health. What happens when we are afraid of what we eat? Ten years ago, I received papers about bulimia and anorexia; now I also hear from students with an anorexic parent, father or mother. Many grow up with confusing signals about food. At a dance festival one summer, a little girl, perhaps five years of age, joined me at the soft-serve yogurt counter of the cafeteria. "Is that low fat?" she asked. "I only eat low fat." Yet food is basic to life, offering daily participation in the cycles and pleasures of the earth. We need fat. It coats our nerves and insulates our bodies from the stresses of our times. Balanced with activity, like an hourlong walk or lively play each day, there is no need to be afraid of drinking whole milk. It's not just the food; it's our attitudes that affect healthy digestion.

providing vertical support for the neck and thorax at the organ level. Swallow or slowly drink a glass of water to feel sensations in this passageway.

The esophagus passes between the lungs and behind the heart, moving the food, now called the *bolus*, on its journey between mouth and stomach. "Heartburn" is a common term for indigestion occurring in this region. The bolus within the esophagus then passes through the breathing diaphragm to the ring of muscular fibers called the *cardiac sphincter* (or gastric sphincter), which generally prohibits digestive gases and food from returning into the esophagus from the stomach (gastric reflux). The diaphragm divides the torso into two cavities, upper and lower, forming the floor for the heart and lungs and the ceiling for the stomach (left side of your body) and liver (right side), with the pancreas nestled front to back between the two. The digestive system is constantly massaged by the rhythmic movement of the diaphragm with each breath. The *stomach* has considerable mobility. Suspended from the adjacent liver and the diaphragm by elastic fibers, it is capable of extending as far down as the bladder after a full meal. The primary function of the stomach is to break down food, using strong digestive juices secreted from the stomach lining.

The *pyloric sphincter* oversees passage of churned food from the stomach to the small intestine. You can generally feel this sphincter by rubbing the area on the stomach side of the torso, about three inches from the base of the sternum, below the left ribs. As a muscular ring, it is harder and denser than surrounding organ tissue. The *small intestine*, commonly known as our "gut," is approximately one inch in diameter and twenty-one feet in length. The upper portion is the *duodenum* (functionally part of the stomach, structurally part of the intestines), followed by the *jejunum* and the *ileum*. Both the liver and pancreas have ducts opening into the duodenum, with secretions (bile and pancreatic enzymes) that help prepare ingested food for absorption. The small intestine intertwines in the cavity made by the frame of the large intestine, by the horizontal bowl of the pelvis, and by the lumbar spine. *Fascia* (common mesentery) contributes to their tight packing, the fatty *greater omentum* (like an apron over the front of the belly organs) offers essential insulation and protection, and the *abdominal muscles* (rectus abdominis, obliques, and transversalis) give support on the front surface and also wrap around to the lumbar vertebrae in back. Blood vessels and nerves arise on the posterior abdominal wall and pass through the tissue to serve the organs. It takes three to ten hours for the food—now called *chyme*—to pass through the small intestines, where 90 percent of the nutrients pass into the blood stream. The remaining 10 percent of absorption occurs in the stomach and *large intestine*. Small capillaries throughout the tract transfer nutrients into the blood. The hepatic portal vein (part of the *hepatic portal system*) delivers the nutrient-rich blood directly to the liver for filtration before cycling it to the heart to be distributed to all the body's cells.

The large intestine is mainly concerned with absorption of water and minerals and some vitamins; it connects to the small intestine in the lower right corner of the pelvic bowl, at the *cecum*, through a sphincter muscle

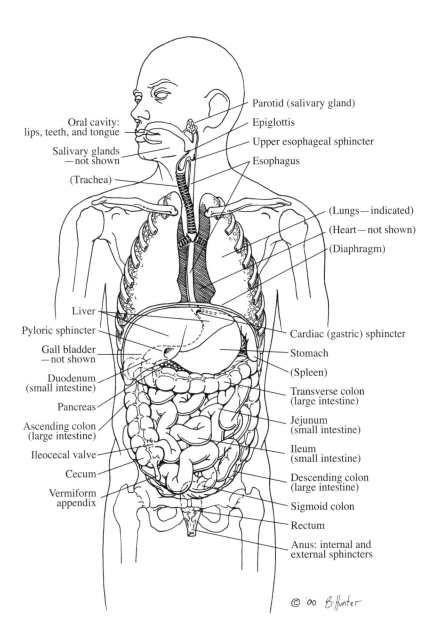

Digestive tract and organs. Anatomical illustration
© 2000 by Brian Hunter.

Parotid (salivary gland)

Epiglottis

Upper esophageal sphincter

Esophagus

Oral cavity:
lips, teeth, and tongue

Salivary glands
—not shown

(Trachea)

(Lungs—indicated)

(Heart—not shown)

(Diaphragm)

Liver

Pyloric sphincter

Cardiac (gastric) sphincter

Gall bladder
—not shown

Stomach

Duodenum
(small intestine)

(Spleen)

Transverse colon
(large intestine)

Pancreas

Jejunum
(small intestine)

Ascending colon
(large intestine)

Ileum
(small intestine)

Ileocecal valve

Descending colon
(large intestine)

Cecum

Sigmoid colon

Vermiform
appendix

Rectum

Anus: internal and
external sphincters

© '00 B. Hunter

called the ileocecal valve. (This tight curve is also where the *vermiform ap-pendix* connects; it has no known function in digestion but contains lymph tissue that contributes to the defense system of the body.) The large *ascend-ing colon* travels up the right side of the pelvis, the *transverse colon* travels horizontally across our belt area, and the *descending colon* travels down the left side. The sharp turn by the left hip is called the *sigmoid colon;* it arches back and becomes the rectum at the middle of the sacrum (S-3). The rectum descends, completing the vertical axis of the digestive tube begun by the esophagus. The *anal canal* and two *anal sphincters* are the last portion of the large intestine and the final exit from the body, expelling the non-nutrient residue of the bolus. Undulations of smooth muscles (peristalsis) everywhere along the tract serve, with gravity, to move food through the system. Tension or indecision at any of the sphincters, lack of tone in smooth muscles of the tract, or constant holding in the abdominal muscles interferes with nutrient flow.

Fluids are absorbed and filtered by the paired *kidneys* during every phase of digestion. Located outside the visceral cavity on the back surface of the body, the kidneys are protected by the lower ribs and are moved during respiration; the right one is slightly lower due to the large size of the liver. The kidneys rest on the front of the iliopsoas muscle, and the *ureter* from each follows the forward diagonal of the psoas fibers, traveling from back body to front body. The two ureters empty their contents into the single *urinary bladder*, located behind the pubic bone within the pelvic bowl. The fluids pool and are stored in the bladder before being released as urine through the *urethra*. As well as playing a role in digestion, kidneys also conserve water and maintain an acid-base balance in the blood, keeping a dynamic equilibrium between excretion and storage of essential fluids.[1] The vital *adrenal glands* rest on top of the kidneys, secreting adrenaline, cortisol, and other chemicals (in dialogue with the pituitary gland) into the bloodstream in times of stress or dynamic activity, which dampens digestive functions and directs blood flow away from the stomach and intestines.

Digestion is closely related to other integrative functions in the body. When we are in the midst of a difficult or new situation, the digestive tract may reflect the uncertainty. For example, if someone is leaving, and you are upset, you may have difficulty taking in the situation and may feel a "lump" in your throat or lack of interest in eating. Sometimes you may be able to take in a difficult experience but resist integrating it into your life, resulting in indigestion—queasiness—or stomach ulcers. Sometimes you can integrate and digest an experience in your life but resist letting go of what is no longer essential from the past, resulting in constipation, diarrhea, or colitis. Not all digestive problems have emotional correlations, but it is worth considering emotional implications if the digestive tract is in turmoil.

By understanding the digestive tract we can visualize food passing through each muscular sphincter, encouraging relaxation and transition. Touch and hands-on massage can be used, particularly with the small and large intestines and around the sphincters, to move food through the tract. Deep breathing or a relaxing afternoon walk can activate tissues and fluids, assisting the dynamic process of nutrient flow in the body. Fluid intake, frequent urination, and regular defecation are essential to remove toxins, although too much fluid (more than eight glasses of water per day) can also flush nutrients from the body.

Rather than considering quiet time as "wasted time," we can see it as the most vital, integrative aspect of our days and nights. Many cultures build time for digestion into the rhythm of the day, with siestas following meals. Dining is also a natural gathering time for families, allowing absorption and exchange. When we plan meetings at lunch and allow our family rituals to be interrupted by phone, television, sports practices, and long work hours, we are sacrificing essential connections such as relaxing, careful listening, and digesting. Try building at least one leisurely meal into your week, and notice its implications for your relationships and your health.

Organs and Alignment

Organs are the contents of the body that facilitate the respiratory, circulatory, digestive, endocrine, and reproductive systems. All organs have volume, weight, and unique "organic" form. Asymmetrical shapes intertwine to fill niches in the internal cavity, creating a three-dimensional puzzle. Some organs have considerable density, such as the blood-filled liver, spleen, and heart. Others are light; the lungs, stomach, gallbladder, and bladder fill and empty. Some are paired—lungs, kidneys, large intestines, ovaries, and testes—balancing our bilateral form, and others are singular and more central, like the bladder, pancreas, and heart. Together, the organs form our internal contents, constantly interacting with the external world around us.

Every organ offers essential structural as well as functional support. The sheer volume of the liver makes it key in alignment of the thorax, the pancreas integrates front and back surfaces, the kidneys fill out the lower back, and the ureters assist in integration of the pelvic girdle—connecting back body with front body through their diagonal axis with the bladder. Organs condense, expand, and rotate internally and are also moved by other tissues. Many organs, such as the pancreas and gonads, have endocrine functions, affecting energy flow through our structure. Organs also can store the emotional history of our bodies; sometimes we hold the heart, hold our breath, hold our anus or intestines as protection, restricting expressive possibilities as well as interfering with physiological functions.

Initiating movement from our organs takes us into deeper levels, allowing expression and mobility at our core. Skilled dancers and athletes, often considered "organic" movers, create outer shape and form through the dimensionality of inner experience. Yet organic movement is part of our everyday lives. As you sit reading this text, wriggle your spine or roll your head forward from the skeleton. Then move your organs to create the same actions, sloshing them side to side, or roll forward with the trachea and esophagus. Each layer provides a different quality of movement, expressive of the whole. Sometimes what we think are skeletal imbalances, like chronically lifted shoulders, actually involve holding in the organs—in this case, the lungs and heart.[2]

In balanced alignment the skeletal system transfers the body's weight to the earth so that the digestive system is free to function without compression or muscular tension. During this process, an expansion occurs through the organs and connective tissues spanning the bones. Continual compression of the front surface (slouching) encourages depression; continual expansion (arching) encourages exhaustion. Dynamic balance involves finding skeletal support and balancing muscle tone. Shift your weight slightly forward (toward the toes), pouring weight into your lungs, heart, and digestive system: "standing in the organs." Then shift your weight back to center (toward your ankles), balancing through the bodies of your vertebrae: "standing in the bones." Now shift farther back, shortening your back muscles and arching your spine: "standing in the spinal cord." Return to center; feel the

Students become discouraged researching nutrition. Studies are often tied to marketing ploys, funded by economic interest groups. Still, there are a few basic truths that always surface: We need water, at least five to eight glasses a day (beyond soda and caffeinated drinks). Vegetables are anticarcinogens; that is, cancer doesn't like vegetables. Organic foods reduce the dangers of pesticides and herbicides in the body. Low-fat diets (low in meat) help us live longer, but some fat is necessary to insulate our nerves. Ideal weight depends upon your body. Beyond that, buy locally; they say fresh produce has the highest nutrient value. Support the family farms in your area; they ensure sustainable food sources, seasonal bounty, and products with natural antibodies for your region. Local organic food is better yet, because you are supporting healthy soil, clean water, and safe working conditions for farmers and harvesters with each purchase. A healthy bioregion ensures nutritious foods.

My mother, a first-grade teacher, offers a simple dining suggestion: Eat something of every color every day—red, green, orange, white, brown, purple. It works. Once a beauty queen, mother has been overweight for the second half of her life. I used to encourage her to lose weight, mainly to protect her arthritic joints from strain. Once she said, "You know, I'm happy. I'm generally in a good mood. I've got lots of energy. I'm not cranky. There are worse things to be than overweight!" And she was right. At eighty-two she teaches swimnastics for senior citizens three times a week, is active from morning to night, and is a warm, positive, and caring person. What's my problem?

difference. In postural alignment, mineral bones to mineral earth, the weight passes through the bodies of the vertebrae, and the responsive organs are free to move, nourish, and inform the body.

Nutrition

We can learn to be afraid of food. Nutrition is the process by which the body takes in and utilizes necessary nutrients from the earth. Human food requirements include carbohydrates, fats, proteins, vitamins, minerals, and water, processed and transported into the bloodstream and carried to every cell. We often encounter contradictory information about nutrient values. We are shown unrealistic views of body image—impossibly thin or strong—and hear troubling stories about the health effects of preservatives, pesticides, herbicides, hormones, and genetically altered food. As we work to afford organic produce, seek out local products in our supermarkets, and consume nutritional supplements to boost or detoxify our systems, the simple pleasures of eating can be laden with complexity.

We also can appreciate our dialogue with the earth through meals. Acknowledging that humans are dependent on other species for nutrient energy, we can honor plants and animals as relatives, supported by air, water, soil, and other humans. We can recognize the contributions of farmers and gardeners, truckers and store owners, cooks and dishwashers, composters and recyclers to our food supply. Muscles and fat are essential body tissues, protecting nerves and insulating and cushioning vital organs and glands; we must consider the appropriate weight and density for our individual structures. Muscle bulk or fat puts stress on other tissues, too little or too much; we need to balance nutrient intake and healthful movement. Attending to our body as it actually is, rather than as we imagine or want it to be, enhances health and vitality and reduces confusion.

Rehabilitating our relationship to food is an important aspect of reconnecting with body and earth. Food, including the process of growing, harvesting, purchasing, preparing, ingesting, and composting, is a cyclic process that reminds us several times daily of our place in the larger picture of things. As we recognize the role of nutrition in our lives, we access the essential pleasure in each of these stages. Yet we all need help in understanding that our foods include nutrients from the soil. Many people are cut off from these life processes. It is helpful to grow plants, like herbs, vegetables, or flowers and develop a relationship with their cycles. If we have rosemary or thyme plants in our kitchens, for example, we participate in providing their requirements for moisture, sun, and nutrient supply, as well as benefiting from their pungent smell, tasty flavor, and oxygen–carbon dioxide exchange in our homes.

Digestion has as much to do with how we eat our food as what we eat. In appreciating food, we participate in one of the basic joys, social rituals, and personal opportunities for community with human, as well as other-than-human, life. Many Americans prefer to be distracted during meals: if we don't pay attention, we can pretend we didn't eat! If we don't grow or hunt our own food, we can believe that death is not part of life. If we don't acknowledge relationship, we can feign autonomy or dominance. Distraction supports destruction of the earth. Digestive and integrative functions in the body need situations of safety and rest. Before eating, we might consider a silent moment for honoring the food—where it came from and all the people involved in growth and preparation and eventual composting and nutrient recycling. Then we can enjoy the aromatic smells and flavors of food, the profound satisfaction of digestion and interaction.

꩜   ꩜   ꩜

Lunch
10 minutes

Seated, after lunch, with a pad of paper:
• Make a list of all the food you ate or drank for lunch.
• Trace the place history of the main ingredients in each food. Visit the kitchen or grocery store and look at some of the containers to see their sources. You may be surprised! Note foods that are local, organic, or of unknown origin.
• Reflect on the relationship of what you are eating to the parent soil. An apple has obvious connections, through the roots of an apple tree, to the nutrient molecules in a particular soil. But even a Snickers bar has its soil history (cocoa beans, milk, peanuts, sugar cane, corn syrup). Be as specific as you can about where each product might have come from: cocoa beans from cocoa trees with roots in the Brazilian soil, milk from cows who eat the grasses in Vermont (local), peanuts from the roots of plants in Georgia, and so on.

Example: (1) sandwich: whole wheat in bread—Nebraska; tomatoes (organic)—California; turkey breast—Tennessee; lettuce—California. (2) chocolate milk: chocolate—Brazil; milk—Vermont (local). (3) apple—Vermont (local). (4) potato chips: potatoes—Idaho; cottonseed oil—Georgia. Include the whole meal. Be as specific as you can. (Note when you have to guess.)

Tracing the digestive tract
10 minutes

Lying in constructive rest or seated, eyes closed:
• Trace your *lips and teeth* with your *tongue*. Relax your jaw and allow the mouth to fall slightly open. Swallow, opening and closing the *throat sphincter* connecting to your esophagus.
• Imagine the *esophagus* traveling behind the trachea, between the lungs and behind the heart, and along the front of your spine through a hole in the diaphragm, to the stomach. Wriggle your torso as if the esophagus were moving you.

In my family, Sunday breakfast was an event: my father would make blueberry pancakes or cheese omelets as a special treat, or mother would take us all to a local park to cook bacon and eggs outdoors over an open fire. So when I had my own family, I continued the breakfast tradition: On Sundays or special occasions, fresh flowers, candles, linen napkins, and crystal glasses set the table for a luscious meal.

When my stepson became a senior in high school, in a particularly rebellious phase, it seemed our family customs were outmoded. One vacation, he invited a special friend to come for a visit. In the morning he asked us to clear our papers off the table and proceeded to make breakfast for two—crystal glasses, candles and all, adding Tabasco sauce and other hot spices to his feast. In some ways this was the most assuring event of the season. Food does nourish—our attitudes, our health, and our relationships to each other.

TO DO

Digestive tract.

• Touch just below the ribs on your left side, and rub your *stomach*. (It may gurgle.) Feel for the *pyloric sphincter,* a ring of muscles between your stomach and small intestine, near your center line, about an inch from the lower left ribs. It may feel rather hard, like a knot, and can be sore.

• Circle your hands around the abdomen, feeling the interweaving of the twenty-one feet of *small intestines,* commonly called the gut.

• Moving your hands to the lower right corner of your pelvis (your appendix), trace the *ascending colon* up the right side of the pelvic bowl, the *transverse colon* traveling horizontally across your belt area, then the *descending colon* traveling down the left side of your pelvis. There is a sharp turn of the *sigmoid colon* (near your left hip socket) before the large intestine arches back toward the sacrum to become the *rectum and anus.*

• Contract and release the *anal sphincters,* just as you contracted and relaxed the throat to swallow. Image the continuity of the digestive tract as one long tube from mouth to anus.

• Visualize an open tract, free from restrictions. Try eating a small piece of bread and tracing its journey.

Place visit: Attention to digestion

Pack a lunch and have a meal at your place. Try to include some local foods available in this season. Take time in the process of eating to breathe, acknowledging the places on the earth that provided this food and the people who participated in growing, harvesting, and preparing it. Then invite your senses to fully participate (smell, sight, taste, touch). Actually taste what you eat; keep your attention focused on the flavors and nutrition provided in each bite (rather than getting distracted by thoughts). Notice when you feel full and nourished. Spend a few minutes in constructive rest, inviting digestion. (Remember, eating is a process not an act.) 20 min. Write about your experience. 10 min. On your walk home, notice if there are any wild edible plants available in your bioregion in this season: berries, wild cucumbers, apples, or nuts.

CROSSING OVER: FIRST VISITS—2

He is walking with bare feet on hot concrete, walking and walking, refusing to rest. We pass a woman, and my father hunches down to imitate her shape—all scrunched over. "She's always like that," he jokes. Back inside the home, another woman sits watching, saying nothing. "She's like you," he tells me.

At dinner he eats every pea, as well as the pile of meat, potatoes, roll, and gravy—methodically cutting, chewing, and swallowing. I sit next to him, wordless, rubbing his back. When everything is done, he begins on the napkin, cutting, chewing. "Spit it out, Dad." He removes the wet wad of paper from his lips like a seed. I ask him if he wants to go for a walk in the outdoor enclosure. "Not out that door," he says.

A new man, Jim, watches us at dinner. He is young, distinguished. Speaking with a British accent, he says to us, "I am impeccably, inutterably, incalculably lost. Can you tell me where I am?" When I answer that people here will help him, he shakes his head and says, "They all try to help, but no one can." I imagine my father, his first night alone. I force myself to look at this man, still without drugs to disguise his pain, aware that he has been left, lost, to find his own way.

The tall, pregnant island woman who delivers the laundered clothes leads my father back to his room. He does what she says. As we sit together on the bed, he confides, "She's one of the good ones. They aren't all like that." I have brought Easter bread from Vermont. He tears off one piece to give to me, one to place by his pillow. I read to him as he roams around the room, taking the coverlet off his roommate's bed, folding it around him. He folds and unfolds and folds. When it is time for me to leave, he has closed his eyes.

Read aloud, or write and read your own story about digestion.

DAY 23

Plants

In our final analysis, man, be he (she) botanist, gardner, or plain Homo Sapiens, is utterly dependent on plants.

Anthony Huxley, *Plant and Planet*

VIEWS

Anne is an extraordinary gardener. Her rural home in southern Vermont reveals a commitment to vegetables and flowers, trees and arbors, fruits and the compost pile, stone walls and ponds. It is worth a trip each July, just to sit amid this bounteous landscape and feel its effects. When I ask her to describe the qualities of a gardener, she demurs. "I didn't set out to be a gardener. I like being outdoors, I like to do physical work, I like dirt, and I like plants; that all added up to gardening." Then she smiles and adds, "Remember, only too many flowers is enough."

Indoors, rooms are enlivened with bouquets in summer and winter, both delicate and flamboyant in scale. Living amid plants is a way of life, as nourishing as a good meal. Her mother, too, is a gardener, with the foresight to create an arboretum preserving open spaces in Dayton, Ohio; then, collaborating with community horticulture organizations, a metro park and river corridors to revive indigenous prairie life along the waterways. Both women's care has an effect on all who pass through the landscape, reminding us that we can create beauty in our lives if we choose.

Plants provide food for every other living organism. They create oxygen that is fundamental to life, store the sun's energy in their tissues, and surround us with habitat and intrinsic beauty. As well as satisfying these needs, plants are used by humans for healing, providing the basis for traditional medicines and the majority of pharmaceutical products we use today. The average Northerner uses plant derivatives from all over the globe, such as exotic hardwoods, rubber, dyes, and coffee. The fields and forests, meadows and grasslands, and our gardens and kitchens all remind us that plants are an essential component of life.

The first life forms on earth were heterotrophs, organisms dependent on an outside source of organic molecules for their energy. In the course of time,

Plum Tree, painting by Philip Buller. Oil on canvas, 48 in. × 42 in. From the collection of Mr. and Mrs. William Jackson.

cells evolved that were able to convert the light of the sun into chemical form. These cells, called autotrophs (self-feeders), created an available energy source for all future organisms. Fossil records from about 3.4 billion years ago show the presence of blue-green algae, a photosynthetic bacterium that could make its own food through the green pigment of chlorophyll.

Unique to the kingdom Plantae are rigid cell walls and subcellular structures containing chlorophyll. Present in every plant species, chlorophyll absorbs the red, orange, and blue spectrum of light, causing them to appear green. Chlorophyll allows organisms to tap the energy of sunlight to make carbohydrates out of carbon dioxide and water, releasing oxygen as a by-product. Chloroplasts are the light receptor cells of photosynthesis, similar to rod cells in the eyes of vertebrates. When light is absent, photosynthesis stops, and the net outflow of oxygen and inflow of carbon dioxide is reversed. Photosynthesis normally occurs in leaves, dependent on available light; in leafless plants, such as cacti and brooms, in short-lived plants like garden annuals, and in plants that grow in light-deficient areas, such as poplar trees under forest canopies, photosynthesis can occur in the stem or trunk.

In the evolutionary story, the surface water became colonized by photosynthetic organisms such as marine algae, requiring sunlight, carbon dioxide, oxygen, and a few minerals for survival. As mineral sources were easily depleted in the ocean, plants inhabited the waters close to land, which were rich in nitrates and minerals from mountain streams. It is postulated that, around four hundred million years ago, green algae gave rise to the first primitive land plants. Although the new habitat provided a ready supply of necessary nutrients, provision for a continuous water supply became an essential factor.

In the transition from sea to shore, several structural changes occurred. Most land plants developed three main organs: a root system to anchor them to a particular place and to gather nutrient minerals and water from the soil; a stem, supporting the photosynthetic parts toward the sun; and leaves, which are the organs of photosynthesis.[1] These structures were interconnected by a vascular system to move food and water through the plant body, working against the pull of gravity. Plants also developed reproductive organs and seeds with specialized coverings for survival in unfavorable conditions. Although plants were now rooted in one place, movement remains part of plant life, turning toward light or opening and closing in response to temperature. Plants also feed, breathe, mate, and produce offspring, interacting with the environment. And the presence of plants does not stop after their deaths; land plants alone create 150 tons of new matter each year, replenishing the earth's nutrient crust.

Mutation and recombination of characteristics have been two essential factors underlying the remarkable diversity of plants on earth. In the plant kingdom (sometimes called the green kingdom) there are twenty classes of plants with chlorophyll as the link. A diagram looks like branches from a stem: bacteria, algae (mosses, liverworts, ferns, coniferous forests, angiosperms, and flowering plants), fungi, and elemental viruses. Of the 380,000

Milkweed surrounds our garden. Its purple, sweet-smelling ball of flowers scents the air in July. Second-growth milkweed feeds the white, yellow, and black caterpillars that transform into monarch butterflies in the late summer. We grow vegetables for us, milkweed for the butterflies. When farming friends came to visit, they said, "Why do you have that weed around your garden? It's a real pest." We stopped to ponder the nature of a weed: it is often defined as a plant that is plentiful, takes over the space, and is considered useless to humans. But our milkweed provides elegant decoration, windbreak, and food for insects. It's not a weed to us!

Hay-scented fern (double tip), collected by Josh Keith.

Living in Tuscany for a summer, I tasted my first plateful of fried squash blossoms. Eating the rich orange flowers heated in oil from local olives after walking through fields of sunflowers, changed my sense of scale, beauty, and culinary delight in a single afternoon. I make these now each summer for guests, blossoms dipped in batter. But today, I just look at the petals spread wide in the sun, not to eat but to taste with my eyes.

Yoga: Tree Pose, Vrksasana.

species of plants known today, about 250,000 are the "higher plants," including flowering plants (angiosperms, "enclosed seeded") and gymnosperms (conifers and cycads, "naked seeded"). The earliest record of flowering plants comes from the pollen of water lilies, which appeared at the very end of the Jurassic period, as dinosaurs were dying out. Today, orchids form one of the largest plant families, with over 35,000 species and blossoms ranging in size from a few millimeters across to extravagant proportions.[2] Both water and land are plentiful with microscopic plants (some unicellular), and numerous larger plants inhabit both shallow seas and the shores.

Trees have a special place in the hearts and lives of humans. They shape and shade our landscape, offer food and habitat for animals, and also secure and replenish the soil. From a purely human perspective, trees have long provided story and myth, fuel for the hearth or stove, wood for homes and tools, limbs for climbing and play, pulp for paper and baskets, bark for healing remedies, oils for cooking and paint, fruits for food and barter, drums for celebrations and spiritual gatherings, and tall trees for canoe rails and the masts of sailing ships. Beyond all this and perhaps most essentially, plants provide beauty and mystery. Their stories are told long after they have died or been cut down. In many cultures, individual or whole species of trees have been experienced as sacred, such as the solitary oak, the giant redwood forest, and the row of cypresses signaling death. Even in our mechanistic culture, many of us slip into the park or woods to hold conversation with our favorite tree.

The energy source in food is stored energy from the sun, processed for us by plants. Peanuts, hamburgers, and chocolate, for example, provide their nourishment through the roots of the peanut plant, the leafy stems and grain crops that feed our beef, and the cocoa trees and sugar cane that combine to sweeten our chocolate treat. What energy there is in food comes directly from the sun via plants. The more direct the route (i.e., the less "processed" the foods), the more available the energy is for human use. The tomato picked ripe from the vine, still warm from the summer sun, offers maximum nutritional value. By eating produce grown locally, organically when possible, we both maximize our nutritional consumption and support regional farmlands. Although we can go to the supermarket and buy food from around the world in almost any season, we might choose instead to participate in the seasonal cycles unique to the place in which we live.

Natural ecosystems are remarkably stable. Humans are dependent on plants, but most of the green kingdom can exist without human involvement. Maintaining a reciprocal relationship with other animals and insects for pollination, reproduction, and carbon dioxide–oxygen exchange, plants would continue to evolve if human presence ceased. Yet beginning with cultivation and farming practices around ten thousand years ago, humans have developed a rather fanatic desire to control plants. In the cause of increasing productivity, size of blossom or fruit, and resistance to disease or insects, humans have been producing strains that cannot survive in the wild. Observation reminds us that diversity provides health in the biosphere. In nature, animals and insects are only occasionally seriously destructive to plants; equilibrium is usually established.

Under cultivation, plants become more susceptible to attack by pests and disease. The overspecialization of genetic strains causes concern to some botanists. Hybridization can be seen as reducing the capacity for self-sustained life and ironically poses a threat to species diversity. An even greater source of concern to the self-sustaining capabilities of plant reproduction are the new technology of "terminator seeds" that cannot reproduce themselves; the cultivated hybrids whose seeds won't breed true (the offspring of those seeds will not necessarily look like their parents); and bioengineered seeds, which are human creations. Some feel it is essential to preserve "pure" strains of seeds for the future of the planet, and therefore they advocate farmers and gardeners to save their heirloom seeds (which will breed true).

Humans are infinitely careless with natural resources. With increasing populations placing huge demands on food supply, we overexploit soil for crops and overgraze, so even the toughest species can't survive, resulting in desertification. In many areas today we are paving once productive fields for parking lots and luxury condominiums. Insistence on perfect lawns creates a monoculture sustained by watering, fertilizers, herbicides, pesticides, and motorized equipment. The desire for green and the ongoing war on weeds (anything different from grass) places unreasonable demands and harmful chemicals in our water supplies, while allowing us to replicate English country houses in such far-flung places as the Florida coast and the Arizona desert, draining essential resources. In fact, weeds can be useful beyond our first impressions. Dandelion greens are a pungent delicacy in a meal, thistle is a source of fiber for weaving, and wild ginger root is an aid to digestion.

Plants are part of natural communities. If we visit biomes such as the rainforests, savannas, woods, deserts, and tundras of the earth, we see that plants define global landscapes. If we experience the great northern forests, the southern Everglades, or the Sonoma desert, we may find that the dense undergrowth, the light-prohibiting canopy, or the barbed spines of the agave plant seem indifferent, even threatening to human presence. It takes time to find our relationship to such an ecosystem if it is not our own, to be comfortable in the wild. There was a reason that American settlers as recently as the 1700s and 1800s cut down every tree in their town or village to bring light to what was considered darkness. Yet, Native American author, Leslie Silko, reminds us that what might seem like wilderness to some is home to others.[4] Native peoples know the details of their landscape like the faces of their relatives, and animals carefully establish their overwintering areas and migration corridors. Our casual reference to these homelands as wilderness can be simply inaccurate.

Hundreds of millions of people in developing countries depend on plants for their traditional medicine as well as for food. Over 80 percent of prescribed medicine around the world is of plant origin. Although only about 5 percent of flowering plants on earth have been examined chemically and

While I was visiting the Lower East Side in New York, my friend Tamar walked me through her neighborhood. She is a dancer who has produced many site-specific works in Manhattan and beyond. The most recent, Demeter's Daughter, was situated in a community garden on this street. Wire fences surround what appears to be an assemblage of rocks, young trees, and struggling grass. "Each day I would collect the needles from the steps outside the garden, to make the area safe for the dancers," she explains. But she and others in the community are proud of this park, lobbying for its preservation. As we converse, block by block, she explains that many of the buildings had been burned in the 1970s; now-vacant lots bring sunlight to exposed soil. The neighborhood is largely Hispanic; families have begun growing gardens, planting vegetables, building small "casitas" for gathering places amid the fruits, vegetables, and flowers. As people come together, each one does his/her part to beautify the whole. Dancing in this neighborhood, using professionals and local teens, had helped her understand this place she has lived in for over twenty years in a new way. As I walk through these streets, amid people's homes and bustling traffic, I feel the importance of plants and of dancing in building community.

During a yoga workshop at Kripalu Institute in Lennox, Massachusetts, I walked down a wooded path to the lake. Observing a variety of plants, I came upon a figure moving quietly, attentively, and I slowed my pace to hers. For several minutes, I assumed her body's alertness—she wasn't looking, I could tell; she was feeling her way with plants. At dinner that evening, unable to resist, I asked what she had been doing. Slightly embarrassed, she explained she was trained in science, as a dietitian and nutritional counselor, "giving me only one dimension of what is true about a plant." Now she is studying herbal remedies, learning to make concentrated tinctures from plants to stimulate the body. "I was listening," she said, "for what the plants have to tell me. It's a complete change in approach; it's the relationship between us which brings the information." Sitting on the lawn after dinner, I felt myself conversing with the grass, the cedar bushes by our sides, the cultivated marigolds. Including their voices with ours, on that summer evening, created a lively exchange.

TO DO

pharmacologically, a quarter of all the world's flowering plants may become extinct within the next fifty years. Even in the United States, many people gather medicinal herbs: comfrey root for coughs, rose hips as a source of vitamin C, and mint leaves for tea. As increased demands for housing and cultivated food decreases land for wild plant and animal life, organizations around the world are working with botanical gardens and arboreta to help save endangered plants.

Each of our yards, garden plots, rooftops, and abandoned lots are a potential source of wildness. Leave a section of lawn unmoved and see what diversity appears, offering habitat for animals; grow an herb garden on your rooftops, like those in Amsterdam; remove concrete from an abandoned lot and plant trees, replenishing oxygen depleted by city traffic; focus on indigenous plants and native grasses suited to particular climates, and help preserve the water table, native birds, and animals; and use traditional seeds to keep healthy strains available for future generations.

Plants maintain the "life-support" systems of the planet. The emergence of green plants 3.4 billion years ago created the oxygen-rich atmosphere suitable for life. They are essential in maintaining the global level of carbon dioxide and water vapor that supports cloud cover, affecting global temperatures. Plants assist in the recycling of water through soil and air, and their decomposing tissues return essential nutrients to the soil. The energy stored in plants is stored energy from the sun. With each breath and meal, we can remember that humans have a role in both preserving and appreciating the plant life around us.

✠   ✠   ✠

Locating a tree through touch (with a partner)
30 minutes

In a wooded area or at your place, the leader with eyes open, the follower with eyes closed:
• As leader, take a few minutes to disorient your eyes-closed partner, by walking in a circle or turning slowly around.
• Supporting your partner, arm around waist, walk slowly until you locate a particular tree. (Note: slow walking seems fast when perceived through touch rather than vision.)
• Place your partner's hands on the tree trunk; let your partner explore through touch.
• Allow time for thorough exploration: roots, ground cover, branches, bark, leaves, etc.
• Lead your partner some distance away and invite him/her to open his/her eyes.
• Allow the follower to locate the tree, with eyes open. (Give a few hints if necessary.) 10 min.
• Trade roles. 10 min. Write about your experience and read aloud to each other. 10 min.

Sunflower with Honey Bee. Photograph by Erik Borg.

Perspectives
Two weeks

- Visit a local farm or garden, organic if possible.
- Visit a local farmers' market and health food stores; note what products are specific to your bioregion. What's in season at the moment?
- Visit a national forest; consider effects of mining, grazing, timber cutting, and off-road vehicles on the land.
- Visit a commercial greenhouse. What are the owners' concerns?
- Read about farming practices throughout the Americas; note the use of pesticides made in the United States, but banned from use in this country (like DDT), imported in the foods you eat.
- Notice lawns and golf courses; visit the toolshed and note the chemicals, fertilizers, and pesticides used to maintain a monoculture of grass; or check the soil—are there earthworms and beetles or other insects aerating the soil, or have they been poisoned?
- Notice indigenous plants; where do you find them, and what are their characteristics specific to your climate and soil?
- Notice non-native plantings; what are their needs (water, fertilizers, etc.)?
- Consider logging and working forests.
- Consider genetically engineered seeds and hormones in animals (such as BST in cows) to increase yields synthetically.

• Consider attitudes about farmers and farming, loggers and logging, and your use of their products.

• Consider that farmers are among the most skilled workers in our country, because they have to know how to do everything, fix everything, and adapt moment by moment to changes in the environment.

• Write about your findings. 2 hrs.

Place visit: Attention to plants

Learn the names of six plants at your place: consider trees, mosses, ferns, grasses, and flowering plants. Notice what characteristics you use for identification: bark, leaves, seeds, stems, flowers, or colors. When you arrive, greet them by name. Close your eyes and keep your awareness on these plants. Can you hear, smell, taste, or feel their presence? Is there another plant that draws your attention? Add vision, and notice how all your senses inform awareness. Pause in open attention, listening to plants. 20 minutes. Write about your experience. 10 min. Next time, bring someone with you for a visit, and introduce your plants by name.

CROSSING OVER: TALKING TO TREES

Today I took the first bite of a crisp, red apple from our land. In less than ten years, some twenty saplings have been planted, with both boys protesting, "Not more trees!"—a magnolia by the pond, apples and pears in the orchard, pines in the wet places. Today, these residents look more familiar on the land than I do, trunks large as a strong man's thighs.

Wild trees populate the soil on their own accord. Black locusts sprout from a latticework of roots just under the ground; pin cherry, dogwood, and poplar move in from the hedgerows. Our neighbor, Harley, was just a boy when his woods were still fields with seedlings emerging. It is easy to see how a farm could become a forest in just fifty years, which has happened twice in this part of New England in the last two hundred years.

Remembering the sweet tartness of our first unblemished fruit, I sit beneath the spreading branches of a Honey Gold. I have talked to trees since childhood, posing questions and listening for their response. Thinking of my father's illness and also of the earth, I ask: "What can I do?" And wait for the answer.

Read aloud, or read and speak your own story about trees.

❧ DAY 24

Fluids

I feel the running tide of my own salted blood.

—Barry Lopez, *Field Notes*

Fluids are connectors. Although we often consider our human body to be dense and solid, fluids comprise approximately 70 to 80 percent of our tissue. All body fluids are the same basic substance—water. This fluid matrix moves continuously throughout the body, changing properties as it cycles through various tissues. Most fluids are located within the cells, between the cells, and in the organs, although different structures also have fluid content, such as the eyes, inner ears, and urogenital track.[1] Fluctuation of fluid proportion within the body is determined by the general state of health and body type. So important are fluids to human life, that a loss of ten pounds of water in the average human body, without replenishment, can be fatal.[2]

Fluids play an essential role in life processes, such as regulating body temperature, cell and organ functions, and elimination of wastes. Eight glasses of water per day is considered the minimum for maintaining optimal fluid supply in our bodies. Most of the water taken in by humans, as with other animals and plants, is given back to the environment. This occurs through evaporation from body surfaces, like sweat on skin; through respiratory surfaces, as we notice by seeing our breath on a cold day; or through urine or

TRANSITIONS

Rivers are teachers. Studying tai ji (t'ai chi) in China years ago, we traveled often by boat, adopting the subtle stance, the mobile center, and the responsive feet of days spent on water. In this way, the lessons were understood in our bodies, before the teaching began. Once, stuck in the Guanghau airport for three days, everyone grew anxious. When we arrived in the beautiful Wu Yi province, our teacher, Chungliang Al Huang, said, "It's all part of the practice: how to flow in life when things are disrupted, as well as the times of ease." Mornings, in city parks, people of all ages moved slowly, fluidly in the landscape. Doing tai ji in these urban environments, we found the river flowing inside us, connecting all that came before with the present moment.

Project Bandaloop, Yosemite National Park. Artistic director, Amelia Rudolph. Dancer, Mark Stuver. Photograph by Steve Schneider.

Rain Girl, woodcut by Karen Murley for poem and dance by Katherine Sanderson.

excrement, facilitated by the kidneys and colon (feces are normally about 75 percent water). In our water-based bodies, the small portion of earth's fresh water that moves through us nourishes and cleanses our tissues, as well as affecting our movement, moods, and states of mind. Although humans play a small part in the water cycle, compared with large-scale evaporation from land and water surfaces, we are part of the larger whole.

Fluids are the transport system of the body, moving such substances as hormones, neuropeptides, and antibodies throughout the structure, for communication, defense, or repair as needed. In this way, fluids integrate the various body systems into an interconnected whole. (Blood and lymph, in fact, are classified as connective tissues.) Fluids move together in rhythmic interaction, reflecting the twenty-four-hour circadian rhythm of night and day, as well as duple (heartbeat), triple (veins), and more complex rhythmic interactions, fluctuating in tempo and accent. Movement is essential for keeping the fluids clean and body tissues nourished, as well as for maintaining balanced internal pressure. Tension or holding in any area of the body can cause restriction of fluid flow, resulting in pain and eventually dysfunction or disease of tissues. Try raising your shoulder girdle up toward your ears, followed by maximum release, to stimulate blood flow in tight shoulder muscles through movement. (This is called puffing a muscle—bringing the origin and insertion together.) Massage the area to pump the fluids through touch.

There are various traditional models for understanding the fluid system, generally focused on studying the parts. In "The Dynamics of Flow: The Fluid System of the Body," Bonnie Bainbridge Cohen explores "the dynamic interrelationships between the fluids as one system." In this view, we can identify seven major fluids in the body, each with its own pump, system of transport, and unique qualities of movement. These types of fluids include cellular, arterial blood, venous blood, interstitial (or tissue fluid), lymph, synovial, and cerebrospinal fluid. The four-chambered heart is the underlying pump for the whole, assisted by the rhythmic movement of breath, by muscular articulations of spine and periphery (walking, bending, twisting, extending, contracting), by the vessels themselves, and by gravity.

Cellular fluid is the fluid within each cell. The chemical composition of a cell is generally about 80 percent water, plus 15 percent protein, 3 percent lipids, 1 percent carbohydrates, and 1 percent nucleic acids and minerals. Characteristics of a cell, the basic structural and functional unit of life, include reproduction, metabolism, and responsiveness to the environment.[3] This underlies the more visible processes of taking in and letting go, moving toward and moving away, and condensing and expanding of all cells and tissues (collections of like cells) in the body. Cellular fluid moves through osmosis from states of higher concentration to those of lower concentration, in constant movement throughout the body, mediated by the semipermeable cell membranes.

The *interstitial* fluid (also called tissue fluid) is found outside the cell membrane. It is clear fluid composed of water and dissolved substances that originate in the blood plasma and leave blood capillaries by diffusion (excluding large blood proteins). Interstitial fluid is generally in a gel-like state, varying from nearly solid to fluid, depending on the mobility of tissues. It is a basic component of connective tissues, helping to connect, support, lubricate, and bind all the body structures together. Continuous movement of fluid from interstitial spaces into blood and lymph capillaries stabilizes fluid volume; restriction causes edema, or swelling.[4] Interstitial fluid moves by hydrostatic pressure exerted by the fluid itself, by diffusion (movement of molecules from areas of higher concentration toward one of lower concentration), by movement of the breathing diaphragm (from pressure changes in the thoracic and abdominal cavities), and by contraction and release of skeletal muscles (within the muscle belly and adjacent tissues). Pumping of interstitial fluid occurs in the fleshy tissue spanning a bone or body cavity rather than at the joints. Flex and release the biceps muscles in your upper arms to feel this pumping sensation, or pretend you are shaking and releasing someone's hand. Notice how the squeezing action of hand and forearm muscles, partnering the bones, moves the interstitial fluids. Try the same action with your feet.

Oxygen-rich *arterial blood* is pumped by the heart to all parts of the body through the arteries (including the nutrient arteries in the core of long bones). In fact, over sixty thousand miles of blood vessels (arteries and veins) provide consistent blood flow, essential for delivery of oxygen and nutrients and removal of waste materials (metabolism), as well as for transport of endocrine secretions (hormones).[5] Arteries are lined with smooth muscle tissue to assist in the pumping action, conducting blood away from the heart. They connect to progressively smaller arterial capillaries (sometimes only one cell in diameter) through which nutrients, gases, and molecular material diffuse into the tissue fluid to cells. In this process, the basic fluid matrix in blood changes to interstitial fluid, and to cellular fluid, reflecting the interconnectedness of the system as a whole.

Red blood cells transport oxygen to body cells and remove carbon dioxide; white blood cells are involved with immunity. Platelets facilitate blood clotting, comprising approximately 8 percent of blood, determining its familiar color. The remaining 92 percent is a liquid matrix called plasma, a clear, pale yellow fluid that consists mostly of water, plus dissolved substances such as nutrients, hormones, respiratory gases, ions, enzymes, and wastes. Human blood has the same saline content as that of our ancient cartilaginous ancestors; the shark, with approximately 1 percent salinity reflects our oceanic heritage. Circulating blood maintains body tissues at appropriate temperatures. Place your hand on your heart or wrist and count this strong, metrical blood pulse in your body, generally around seventy to eighty beats per minute, depending on body type, health, and state of arousal. Jog in place or do a few push-ups to stimulate arterial blood.

Venous blood travels through the veins, returning 90 percent of the deoxygenated blood from the tissues toward the heart (the remaining 10 percent returns through the lymph vessels). Veins have fewer muscle fibers than arteries have, and venous blood relies on breathing and other muscular

I teach "the fluids" a few weeks before graduation, focusing on transitions. In the body, all fluids are essentially one. Blood, lymph, cerebrospinal fluid, cellular fluid, and interstitial fluid are basically water, with added components for their unique roles in the body. In this work we differentiate each as a distinct system, with its own pump, transport system, and resulting quality. We begin with the blood. Listening to our hearts, the muscular pump for arterial blood, we engage cardiovascular activities so popular in our culture: running, aerobics, and games. As we slap each other on the back and call out greetings, the room grows noisy, interactive, extroverted.

Moving through all seven fluids, we experience the full range of possibilities within our structures. Some feel awkward; some are familiar. There is no right, no wrong; no valuing of one state over another. We are recognizing and supporting the range inherent in our bodies. And in the process we learn to make a transition through our fluids, instead of tensing our muscles when we come to a moment of change. Transitions are important; they move us through our lives.

movements of the body, particularly those of the hands and feet, to help move the fluids back toward the heart. If you listen to veins through a stethoscope, you will hear a slow triplet rhythm, calming and quieting like a waltz or waves on a beach. Sometimes their thin walls distend, as venous blood pools, waiting for sufficient pumping action. Valves keep the deoxygenated blood from traveling in reverse. Swing your arm in a pendular movement to stimulate the venous fluid, returning blood to the heart.

The *lymphatic fluid* is clear tissue fluid that has entered a lymph capillary and become part of the defense system of the body. Approximately 10 percent of blood fluid returns to the heart through the lymph vessels for filtering in the lymph nodes. Lymph vessels are present throughout the body, except in the central nervous system. Lymph cells, including the lymphocytes and macrophages that are involved in immune surveillance, are produced in red bone marrow and within lymph nodes. Concentrations of lymph nodes occur in the tonsils, neck, armpits, and groin area, as well as in lymphatic organs, including the thymus and spleen. Large protein molecules and foreign substances are transported to these tissues for processing; then the freshly cleaned lymph fluid is carried by lymph vessels to collecting ducts, entering the venous system just before blood returns to the heart. Although there is some smooth muscle tissue in large lymph vessels, lymph is primarily pumped by muscular movement of the whole body. Try clear, precise peripheral actions, like shooting a bow and arrow or gymnastics, to stimulate lymph.

The *cerebrospinal fluid* (CSF) bathes, nourishes, and provides shock absorption for the brain and spinal cord. This fluid also may travel out the peripheral nerves and empty into the interstitial fluid throughout the body.[6] Produced primarily in the lateral ventricles within the brain, the CSF completely surrounds the brain and spinal cord, then passes between the membranes covering the brain (meninges) and is reabsorbed into the veins, which return the fluid to the heart. The CSF is secreted and reabsorbed continuously, maintaining continuity of pressure, and is pumped by movement of the pelvis and spine. The CSF has a rhythmic cycle slower than the heart, moving through the brain and spine at approximately thirteen cycles per minute.

Synovial fluid is secreted by the synovial membrane of joints. A clear fluid, about the consistency of raw egg whites, it nourishes and cushions bone ends and reduces friction in movement. When we first wake up (or try to move after prolonged periods of sitting), synovial fluid can be in a gel state, resulting in a feeling of stiffness. As synovial fluid is warmed by the movement of bones, the viscosity of the fluid lessens; thus, joint movement flows more easily as you "warm up." Movement also stimulates the synovial membrane to increase production of this fluid; injury can cause too much fluid to accumulate in the joint capsule, causing swelling, dangerous to the integrity of tendons and ligaments.

Change of fluids from one state to another, such as the movement from interstitial to synovial fluid and back, is essential to healthy functioning of our fluid bodies. All fluids are one, with the capacity of changing location and function. Because fluids are the transportation system of the body, integrating the various parts, most movements reflect a blend of fluid states, comparable to the interconnected rhythms of water in the world around us. In fact, all dynamics found in water are also present in the body fluids, such as pooling, pouring, crashing, flowing, draining, surging, and stagnating. Holding in any tissue affects fluid flow through the whole.

⊰⊱   ⊰⊱   ⊰⊱

Moving with the fluids* (alone or with a group)
45 minutes

Standing, eyes open or closed:
• Stimulate the pump for arterial blood: your heart. Begin with jumping jacks, running quickly through space, and interacting with people and place. Follow impulses for bloodful, energized movement; keep your eyes engaged, seeing what is around you. Arterial blood helps energize your body.
• When you finally get tired, stimulate the pump for venous blood: the periphery. Swing an arm, leg, or your head, in a pendular motion. Feel release of control as you follow the momentum wherever it leads you. Drop your neck and eyes, and take a ride. Imagine you are on a beach simply "going with the flow." Venous flow helps relax your body.
• Stimulate the pump for lymph: the dynamic pull between periphery and center. Practice fencing, shooting a bow and arrow, hitting a tennis ball, or leaping precisely from one rock to another. Any movement of periphery, focusing on accuracy and correctness, engages lymph. Keep your eyes on the target, with a direct, clear gaze. Lymph underlies clarity of goals and boundaries.
• Stimulate the pump for synovial fluids: the movement of joints and bone ends. Jiggle the joints, moving the bone ends to bring up the synovial fluid in the joint capsules. Scan your body to include all movable joints. Let the eyes release in their sockets, like dancing to funky music. Synovial fluid helps with playfulness and relaxation.
• Stimulate the pump for cerebrospinal fluid: movement of the pelvis and spine. Stand tall and lift your arms slowly as you gaze at the distant horizon (it feels like tai ji). Cerebrospinal fluid engages an expansive and quiet mind.
• Stimulate the pump for interstitial fluids (tissue fluid): the movement of your muscles. Contract and release the bodies of the muscles, moving the tissue fluid. Like a giant cat, sensuous and alive, striding across the plains, stimulate sensations by pulsing muscle. You will feel hot, visceral, flooded by sensation; let your eyes move with the body, wherever it goes. Interstitial fluid supports sensuality and heat in the body.
• Stimulate the pump for cellular fluid: the cell membrane and surrounding tissues. Engage a body scan or meditate on breath. Maintain a calm focus.

*Exercise drawn from the work of Bonnie Bainbridge Cohen, founder and educational director of the School for Body-Mind Centering®. For more information, read "The Dynamics of Flow: The Fluid System of the Body" in Cohen's book *Sensing Feeling and Action.*

TO DO

OF SPECIAL INTEREST: MOVING THE FLUIDS

Paddling on a three-day canoe trip on beautiful Nicatous Lake in Maine, I think of fluids. When we leave the dock, the water is calm, allowing the ease and heightened awareness of cellular fluid—everything is clear. As the wind picks up, arterial blood matches the pounding of waves, causing me to interact and pay attention. As the weather becomes wild, blowing us up-lake, precision paddling is essential, engaging the lymph as I place each stroke into the wave effectively for maximum momentum without flipping the boat.

Resting in the lee of the island, out of the wind, our canoes bob up and down on calmer water. It all seems humorous, as we tell jokes and relax into synovial fluid, recovering from the strain. Reengaging the fierceness of wind and waves, we paddle the long distance to our campsite. Physicality dominates all thought, as I surrender to the animal pulse of muscles, energizing sweat, and the kinesthetic rhythm of interstitial fluid, sensual and hot.

Collapsing by the campfire, listening to the lapping of waves against sandy shore, we indulge the restful ease of venous blood, lulling us to sleep. The next day, paddling under bright blue skies with the crystal-clear Maine light glinting off serene water, the lake inspires larger connections. Our human bodies become a vertical axis linking sky and earth through the cerebrospinal fluid, evoking the spiritual calm of being one with water. The same trip, the same lake, the same body contains the whole. All fluids are one.

Photograph by Erik Borg.

Move and view the world around you with cellular awareness. Cellular awareness underlies attention.

• Imagine the transitional fluids moving from cellular to interstitial, from one state to the next, through the cell membranes. Move from the image of transitional fluid.

• Return to awareness of the fluid matrix, which includes the whole.

• Pause, in open attention.

Place visit: Attention to fluids

Walking to your place, notice which fluids are engaged. Are you in a rush (arterial)? Are you trying to be on time and get this assignment done correctly (lymph)? Are you meandering (venous)? Don't judge, just notice the fluid states that are with you at this moment. When you arrive, experience your place in different fluid states. Begin with what is easiest for you at this moment. Try changing—through the fluids—to another state. For example, begin in constructive rest (cellular) and transition to looking specifically at details (lymph). Get up and do things: climb the tree, roll in the grass, jump off the rock, interacting with place (arterial). Transition to jiggling the bones, bringing up playfulness (synovial fluid). What have you missed? Usually what we can't remember is the one hard for us on this particular day, but it changes! Is your experience of place affected as you move through the fluids? 20 min. Write with awareness of the different fluids, and notice the qualities that emerge, such as accuracy, passion, sensuosity, stream of consciousness, mystery, calm, and playful fun. Read aloud. 10 min.

CROSSING OVER: SKIN

We have the same skin. We get blisters from not speaking, ulcers from caring too much. Now his skin is dry, transparent. If they grab him too hard to make him do as they say, his skin comes off in their hands. Stepping back, I look at wrinkles, bruises, blood; then chair, window, bed. Opening my palms to stale air, I practice distance.

I begin to rub his swollen feet, trying to feel a pulse. "Not so hard," he warns me with a father smile. "You'll hurt me." That's his old voice, familiar. He lets me circle toes, find movement in ankles, rub his knees. His eyes close. He grasps a can of Diet Coke, swigging deeply. A basketball game is on his roommate's TV, sound turned down. "Dunk it," he calls out softly, then sleeps. As I stroke his right wrist, I feel nerves let go. Jerks twitch through rigid spine all the way out to his feet. He is trying so hard to hold on. I say, "It's OK, Dad. It's OK with me if you need to leave." I put my hand on his heart, hear his sigh. "Go, before it gets too hard." At his head, I touch the taut muscles of neck, the peeling flesh of scalp. "I'm so sorry we had to leave you here in this home away from home." Holding the other wrist, "Take your own time." At the toes once again, I feel his reflex to press against my palms. I want his feet to feel the ground. It is my desire. I don't want him to fall. To break bones. To be in a wheelchair. But I know he will.

To say good-bye, I kiss him on the right cheek. He takes his hand, reaches over to mine, and pats it softly. A tear forms in his left eye. I don't want to go, and I look in the hallway for a reason to stay. The gardenia I brought as a present sits in a glass of water behind a white plastic chair. I carry it to his bedstand and say again. "It's OK, Dad. Do what you need to do." Kissing his left cheek, I breathe deeply of the familiar flesh and bones and breath of my father living. It is still a sweet smell, not yet of death.

Read aloud, or write and read your own story about fluids and transitions.

⊰ DAY 25

Water

In silence, I pulled through the water and saw how a river appeared through rolling fog and emptied into the lake.

—Linda Hogan, *Solar Storms*

Water is essential for life. The first organisms emerged from the oceans, and all living species are dependent on water for survival to keep them from drying up. Water also supplies nutrients and removes waste materials, provides habitat, monitors temperature internally and externally, and offers a fluid medium for reproduction and travel. Water is the major component of the planet's surface, sculpting and redistributing the soil. It also regulates temperature through the water cycle and links places within continents and continents to each other. We can be standing on a dusty plain without a river or ocean in sight and still be surrounded by water. Every cloud, tree, blade of grass, insect, and animal in view, as well as one's own body, is mostly water. Water connects the whole.

Night Flight, Millerton, N.Y. Photograph by Peter Schlessinger.

A photograph of our planet seen from space reveals that the earth's surface is about 71 percent water, roughly the same percentage as fluids in the human body. Most of the water on the surface of the earth is in the oceans, with a salinity of approximately thirty-five parts per thousand, comparable to the salinity of human blood. Only 3 percent of water on the earth is fresh water, suitable for drinking; and of this, 2 percent is locked in polar icecaps and glaciers. Of the remaining 1 percent of earth's water that is fresh, 0.9 percent is liquid (in ground water, lakes, ponds, rivers, and streams) and 0.1 percent is in gaseous form as water vapor. (One way to visualize these proportions is to think of a pint water bottle representing water on the planet; one teaspoonful represents fresh water, and one-half drop is available fresh water.) Surface water, however, constitutes only about 10 percent of the total amount of water in the earth. The majority is chemically bonded to minerals in the mantle.[1]

Water, two hydrogen atoms and one oxygen atom bonded together, has the capacity to transform into gas, liquid, and solid states, depending upon temperature. As a gas, water vapor in the atmosphere travels around the globe in the troposphere (zero to ten miles from the surface), condensing around particles into water droplets, forming clouds, and cleansing the air through precipitation. Water vapor also exerts what has been called a greenhouse effect, absorbing much of the sunlight striking it from above and also much of the radiation re-emitted by the earth. The absorbed radiation warms the atmosphere, which in turn warms the earth's surface. As a liquid, water has surface tension, low viscosity, and the ability to hold and store heat, helping to maintain the earth's temperature within a range favorable to life. As a solid, the polar ice caps and glaciers stabilize the effects of weather, moderating world temperature. Thus, water in all its forms plays an integral role in creating a habitable planet.

Earth's water is more than 3.8 billion years old, with the exception of a very small percentage that has been released from the mantle in more recent volcanic activity. As we remember from our study of underlying patterns, the geological record shows that the planet is approximately 4.6 billion years old. Sedimentary rocks (those deposited by water) have been discovered that are 3.8 billion years old, marking the earliest recorded water. There are two main theories explaining water's origins. One is that the earth's water was formed from oxygen and hydrogen released from the volcanic activities of rock. Another postulates that a type of meteorite, carbonaceous chondrite, contained a substantial amount of water in its mineral structure—20 percent by volume. As these meteorites bombarded the protoplanet earth, they produced enough heat to release the water from its mineral structure. In vaporous form, this water became part of the earth's earliest atmosphere.

The total amount of water in the earth's system has changed very little since then. Both the massive volcanic eruptions and the bombardments by meteorites, two possible origins of water, have stopped. A small amount of water continues to be added from the earth's interior through volcanic eruptions, but an equal amount escapes into the atmosphere, balancing the total

INSIDE STORIES

"There is no water at the place I visit," a disgruntled student writes. He is responding to an assignment to notice water during his weekly site visit. "I didn't choose a place near the river or a pond, so I can't do this assignment." And this is the point, I think to myself: to recognize how much water is part of our lives, whether we can see it or not. What about the clouds, I respond, the water underground, in grass and trees, the squirrel, your own body? Water is everywhere, even in the desert. We just have to know how to look.

Deer Canyon is a tribute to water. Standing on an outcrop for a last sun-filled look at the Colorado River below, we turn toward the damp hidden world of this side canyon and continue our climb. Striations of sediment line its undulating walls. Following the sure footsteps of our young guide, moving confidently along the increasingly precipitous path, I extend my left hand for support. Stabilized by rock once scoured and compressed by water, we ascend slowly. White palm prints mark the underledges, a thousand-year-old message, perhaps from the Anasazi who passed here before; I slip my hand in theirs, responding.

As the path narrows, I avoid looking up toward the rim, or down, at the serpentine water below, resisting a kind of panic of falling. Instead, I focus closely on the details of rock, the cool air on the skin of my face, the touch of each foot to earth as they find their hold around an outbulge of rock. This is all preparation, it seems, for the ease of passage that follows. A horizontal meander of water over rock leads us to a pool where others in our group are already gathered. A weathered and wizened tree graces this small pool; moss, ferns, and the sounds of falling water fill every molecule of the space. I squat amid the roots, then slide my body into the torrents, submerging and reappearing behind a screen of falling water. In this moss-lined cave, I find myself looking from the inside, rather than the outside, as inhabitant rather than visitor to Deer Canyon, for this moment.

amount of water on earth. Thus, the water we drink today is the same water, recycled for billions of years through evaporation and precipitation, as that consumed by the dinosaurs and our ancient hominid ancestors.

Movement is a component of water. The fastest movement of water molecules occurs at higher temperatures in water vapor; the slowest, in frozen ice caps. There is a constant exchange between the earth's atmosphere (air) and the hydrosphere (water). The latter is composed of oceans, lakes, rivers, groundwater, ice, snow, and the relatively small amount of water vapor that exists in the air. Currents of air provide an essential medium for water to be distributed around the globe; oceans store heat, while wind currents redistribute that heat in the latent energy of water vapor. The movement of water is also a major geological force, shaping land and redistributing materials through erosion and deposition.

The *hydrologic cycle*, powered by energy from the sun and by gravity, is a circulating process of evaporation, condensation, infiltration, and evaporation. Water (a) evaporates from the land, water, and organisms and enters the atmosphere as water vapor; (b) condenses and is precipitated to earth's surfaces as rain, snow, sleet, and hail; (c) moves over land by runoff into river and lakes, and eventually is absorbed into the ground through infiltration or returns to the seas; and (d) once again enters the atmosphere through evaporation from exposed water or soil, from plants through transpiration, and from animals through sweat, urine, and breath. Although the total amount of water on the earth remains about the same from year to year, the water cycle moves it from place to place.

The renewing functions and natural purification of water through the hydrologic cycle takes time. The average water molecule resides in the air for ten days, but it can take decades for water to filter down to the water table to become accessible as drinking water. Each tiny water molecule spends approximately 351 years on land and 3,200 years in the ocean to make one complete trip through the hydrologic cycle. The water in one tree is replenished thousands of times as the tree grows. The one hundred pounds of water in an adult person's body is replaced seventeen times a year, and the oceans' water is renewed by streams every forty thousand years.[2]

Increased paving of the earth's surface, including sidewalks, roads, airports, buildings, and parking lots, blocks absorption capacities of the soil. This, in addition to overcutting forests, overgrazing, and inefficient farming practices, causes erosion. Vast drains on water resources from increased population and from irresponsible irrigation of lawns and crops are altering the natural water cycle and endangering the earth. In contrast, the preservation of wetlands and marshes offers valuable resources for water purification, providing natural holding areas for overflow and allowing slow absorption of water into the soil.

A *watershed* provides a way of looking at land from the perspective of water. Each watershed comprises the area drained, from the highest topography to the lowest, by a network of creeks, streams, lakes, and rivers, creating natural borders and diverse habitat. When we use watersheds rather than political boundaries as a lens for looking at land, we develop a dynamic picture of the importance of water to place. Essential questions emerge: Where are the highest points in our watershed, and what are the contours of the land that affect drainage into and through the lowlands? In what directions do the rivers flow, where are their sources, and where do they join other creeks, rivers, and lakes? How is the water transported to the sea, and what is added to it in the journey? What are the various soil profiles, and how is the water absorbed into the water table? What holds soil in place—trees, shrubs, grass? Where are the urban centers, sewage treatment plants, factories, farmlands? What is the role of wetlands as essential components of the whole? All of these questions help to define watershed characteristics and boundaries within bioregions, on a scale from local to global.

The earth's water is often explained as interacting compartments in which water is stored for short or long periods. We can consider the oceans; lakes and ponds; rivers and streams; ground water; glaciers, ice caps, and snow fields; and clouds, mist, and fog as components in this process:

There are four major bodies of water classified as *oceans*, making up 97 percent of the earth's surface water. They are the Atlantic, Pacific, Indian, and Arctic Oceans. While the ocean water has existed from early in geological history, the ocean basins are fairly young due to movement of tectonic plates. Almost all of the current basins are less than 250 million years old, influenced by the gradual separation of Pangaea into various continents, creating new ocean floors. In this process, Panthalassa—"all seas"—was divided into distinct oceans. The salinity of the ocean is caused by the wearing down of continental rock and is a major factor (along with temperature) in determining the constant circulation of ocean waters. The saltier water is denser than fresh water entering from rivers and streams; as surface fresh water is blown by prevailing winds, the salt water rises from the depths to replace it.[3]

The ocean floor has elegant and extreme contours, including vast plains and the highest mountains and deepest canyons on the earth. As with all edge habitats, the juncture between land and sea creates a dynamic ecotone, called the intertidal zone, offering a plethora of life. Oceans are often nutrient-rich near continents due to shallower and warmer waters on continental shelves, providing habitat and food for many species.[4] The rhythm of the tides, appreciated by beach walkers and fishermen worldwide, reflects the repetitive pull between the centrifugal force of the spinning earth and the gravitational forces of the sun and moon.[5] Increased human inhabitation of our coastal edges challenges this diversity of life, threatening primary production (the tiniest microorganisms at the base of the food chain, which support all higher aquatic organisms); important nesting grounds of birds; the habitat for rare, delicate grasses; and the ever changing terrain of our barrier islands.

The presence of an *ice or snow cover*, in relation to living organisms, is an essential factor in nature. Snow creates a blanket of protection for plants and animals overwintering in harsh climates, insulating the surface of the ground by trapping air, like a sleeping bag maintaining warmth around a

To paddle up the Connecticut River, we drive north from Guildhall, Vermont. Launching our canoe into the current, we note that the bank on the New Hampshire side is stripped of all trees. You can lose an acre of soil in a year in flood season, my husband reminds me, without roots to hold the soil. Yet there is no industry north all the way to the headwaters near the Canadian border. Trout, picky about water quality, have lured fly fishermen to these banks. We pass two now casting their dry flies upstream, letting them drift down with the flow and hoping that brown trout (introduced from Germany), rainbows (from Oregon and California), or diminished native brookies will bite. We too move upstream against the mild current, as red leaves drift past on the water. Our return will offer an easy ride. Traveling with the water in late afternoon, all we will have to do is steer.

Continuum is a movement form based on the fluid aspects of the body. Originally developed by Emilie Conrad and taught with Susan Harper, it helps us dissolve fixed concepts of limitations in the body. Exploring tiny undulations, wave motion, micromovements, and breath, we move as if the body is water, in water. It has been twice before, in the womb and in our evolutionary past. Engaging these patterns that existed before our personalities were shaped, Continuum workshops are invigorating, inspiring, and fun. Fluids seem to bring out our lighter nature, dissolve rigid concepts, and help us connect. Walking after a session, you feel the fluidity of your once solid form. It is easy to retain fixed models of who we are or who we need to be. As we attend to the wave motions happening inside and outside our bodies, we dissolve, instead, into the timelessness of the moment.

human body. Ice and snow also provide essential water resources in spring melt, ensuring moisture for new growth, and sufficient groundwater to replenish streams, lakes, and the water table (the level to which a well would fill with water). Polar ice stabilizes the effects of the weather and helps us to predict future movement of glaciers. Ice and snow cover also play an essential role in maintaining the earth's temperature, creating a large reflective surface that deflects the sun's light and heat back into the atmosphere.

Lakes and ponds (and man-made reservoirs) are natural holding areas for water and vital resources for humans and other organisms. Formed by glacial potholes and surface indentations, they are fed by natural springs, streams, rivers, and rainfall. In some areas they provide the water supply for major cities—such as Boston, which draws its water from the Quabbin Reservoir sixty miles to the west. *Swamps, marshes, and other wetlands* play a vital role in the hydrologic cycle and are home to innumerable species of plants and animals. In fact, swamps and bogs are the most productive ecosystems on the earth, surpassing even tropical rainforests. Their dense vegetation stabilizes soil and holds back surface runoff, creating a steady year-round flow. They offer habitat for migrating birds; for amphibians such as frogs, salamanders, and toads; for reptiles, including turtles and snakes; and for mammals such as beavers, otters, and muskrats. They also act as buffers between human disturbance areas (roads, development, and agricultural fields) and water bodies, filtering pollutants from runoff. The disturbance of wetlands results in flooding and erosion, which have an extremely harmful effect on biological diversity and productivity.

Rivers and streams are connectors, providing links to other bodies of water. As unique running-water (lotic) habitat, they support a diversity of freshwater communities and are important components of watersheds. Rivers and streams are formed from precipitation that runs off the land surface, drawn by gravity back to the oceans, rather than being absorbed or evaporating. The flow of water is influenced by the contours of the land; speed varies according to environmental conditions and cycles and determines the composition of the river bottom, whether rubble, pebbles, sand, or mud.

Rivers move over time in response to changes in environmental conditions. Building dams, dikes, and houses in floodplains generally reflects ignorance of the fragile ecological web of life supported by rivers, as well as a misperception that rivers are static entities that remain the same in all conditions. Human structures affect the banks, or riparian zone, which provide habitat for stabilizing vegetation and diversity of animal life, both on land and in water. The rerouting and straightening of rivers' courses can affect water temperatures and river currents, causing buildup of sediment that can destroy fish and other aquatic populations, as well as causing more flooding downstream by speeding up the flow.

The *groundwater* provides essential drinking water resources for humans and other living organisms. Within one kilometer of the earth's surface, there is more fresh water than thirty times the volume in all the lakes, rivers, and reservoirs on the surface.[6] Developments such as urbanization and road building are preventing the replenishment of this groundwater, and some is contaminated by pollutants, such as gasoline from underground storage tanks seeping into drinking wells, careless disposal of industrial poisons, or pesticides and fertilizers draining from fields. When considering our drinking water, it is useful to remember that what we put in the air and soil eventually ends up in the water and, consequently, in us.

The *water vapor* in the atmosphere forms our clouds. The amount of water vapor the air can hold increases with higher temperatures. When the air becomes saturated with water vapor (100 percent humidity) and then is blown to an area of lower temperature, the air can no longer hold the water and condensation occurs, creating precipitation such as rain or snow.[7] Clouds carry latent heat energy around the globe as part of our weather systems, thus affecting humidity levels. Fresh water is distributed from the atmosphere over land masses through the movement of water vapor. For example, clouds formed over a lake are blown over the land and drop their precipitation. Water vapor is also the reflective element of a rainbow, as sunlight shines through water-saturated air.

Water molecules (two hydrogen atoms and one oxygen atom) give form and support to many cells and tissues in plants and animals. These molecules are essential to metabolism; the nutrients, materials used by almost all living organisms, must be dissolved in water, which flows throughout the structure and integrates the whole. For example, in humans, adrenaline secreted by the adrenal glands travels through the bloodstream to activate specific organs throughout the body.

One third of the world's population lacks safe drinking water and sanitation. Yet we use three to four gallons of fresh water every time we flush the toilet. Two quarts of water per day per person are optimal for survival, whereas the average Vermont resident consumes 302 gallons of water a day—excessive but the lowest of any state in the United States (66.6 percent of the fresh water used by humans goes toward irrigating crops). We dump harmful chemicals and trash into our limited freshwater resources, build subdivisions without adequate water supplies, and spray our lawns and fill our swimming pools as though the earth offers an endless supply of water for our personal use. Meanwhile, fish are suffering from lack of good habitat (a recent study shows that 40 percent of the fish caught in the United States are too polluted to eat), bottled water has become big business, the exhaust from our motorboats leaves a black fluffy layer on the bottoms of swimming lakes, and streams exhibit abnormal algae growth from phosphate pollution.[8] Yet, we can work individually and with the many regional or national organizations now formed to care for our water sources and resources. We can refrain from use of chemicals and pesticides in our lawns, use more efficient motors, and help stabilize soil along easily eroded stream banks by preserving or replacing indigenous trees and shrubs.

Water connects us to the rhythms of the natural world; it is a gathering place for animals, people, and plants. As we observe crashing waves, tumbling waterfalls, or a drop of rain running down a pane of glass, our bodies respond. We are affected by water through our senses: through the warm temperature of a shower on skin, the forceful movement of swimming in the

Drawing by Jonah Keith, age twelve.

sea, the subtle taste of drinking water, the evocative smell of rain or damp earth, and the dramatic sound of a thunderstorm. Water has long played a role in healing, including the hot springs and seasides that offer rest and recovery, the baths or showers that relieve tension, and the daily requirement for water in our diet. In presence or absence, water defines life on earth.

Water is part of our language and our imagination. In the psyche, the oceans mirror the unconscious as vast sources, both destructive and generative. Water is a common image in dreams, part of the universal imagery of the unconscious. Creation myths tell of animals diving to the depths of water to bring up a handful of mud to shape humans; or a boat, crossing the water, as an image of death. It is part of our metaphorical language: the river of life, the ocean of emotions, and the stream of consciousness. Fluid, flowing, watery, wet, all creative forms draw on water for inspiration, to evoke the imagination and the senses. Without the unique properties of water, life on this planet would not be possible.

⇗   ⇗   ⇗

TO DO

Reading the signs: 10 questions to ask a river*
2–4 hours

Using field guides and notes, visit and write about a river near your place. Use this conversation with the river to guide your reflections.

1. Where do you start and where do you end?
2. What kind of vegetation is on your banks?
3. Who or what lives in you, on you, or near you?
4. Do you have oxbows, an estuary?
5. Who puts things into you?
6. Who joins you on your journey; whom do you join?
7. What makes up your bed?
8. What is your relationship to humans?
9. How are you affected by seasons and storms?
10. What is your aesthetic profile—color, smell, taste, texture, temperature, rhythm, relationship to the landscape?

Tai ji (t'ai chi): Receptive stance
10 minutes

Standing, indoors or outdoors
• Stand in a "tree hug" position. This is the basic stance for beginning tai ji: feet slightly apart, knees bent, spine fluidly vertical, arms creating a relaxed circle in front of your shoulders, and eyes receptive. This alignment includes the empty space in front of the body, where the *chi* energy cycles, traveling down into the earth, up and around your body, and into this potent space.

* Exercise drawn from Jeff Meyers, naturalist and executive director of Vermont River Conservancy.

Garden sculpture, by Herb Ferris. Manzanita basalt.

Tap your belly a few times, stimulating the *chi*, the essential energy that integrates body and earth but is concentrated in the belly, your *tant'ien*. Breathe fully.

• In receptive stance, imagine you are standing on a boat in water. Your ankles, legs, and spine are responsive to subtle shifts. Keep your energy flowing and stay in this position for five minutes, breathing fully. If you lose the energy flow, the muscles will tire; if the energy moves through the whole body, supported by the earth, the task is easy.

• Pause and relax. Take a walk; imagine water moving under and inside your body.

Repeat several times over the course of a week. If you are interested in this exercise, read *Embrace Tiger, Return to Mountain* by Chungliang Al Huang or study Qi Gong, the ancient Chinese healing art that is the basis for tai ji.

Place visit: Attention to water

Bring a bottle filled with drinking water. Sit comfortably and focus your attention on water; bring all your senses to meet the water at your place. Every few minutes, take a few sips of water, "waking up" to water, noticing the transition from external to internal. Consider how every living thing you see at your place requires water for survival. Continue until all the water is gone from the container, but is inside you. 20 min. Write about your experience. 10 min.

CROSSING OVER:
RETURNS

The last time I saw my father, in an Alzheimer's home in Florida, he said, "Well hello, honey," wrote "father" on his placemat, spelled out his name. Pretending we could leave, I rolled his wheelchair past the admissions desk to the parking lot, watched cars pass us by.

My last image was of him tied to the bed. It was noon and there he was, seated on the edge, pale blue dressing gown, bare feet. I handed him an *American Artist* magazine, and he assumed a familiar posture: legs crossed; fingers tracing paintings of mountains, birds; gentle hands turning the pages calmly as I left.

There is nothing boring about death. Three months later, I kneel on the gold carpet at my mother's house, holding a plastic bag filled with ashes, sealed and numbered. I shift its contents in my hands, shards crumble under my touch. I need to feel its weight, before dispersion into air, water, soil; tiny specks scattered into the ocean where all this began.

Read aloud, or write and read your own story about water.

 **III**

Connections

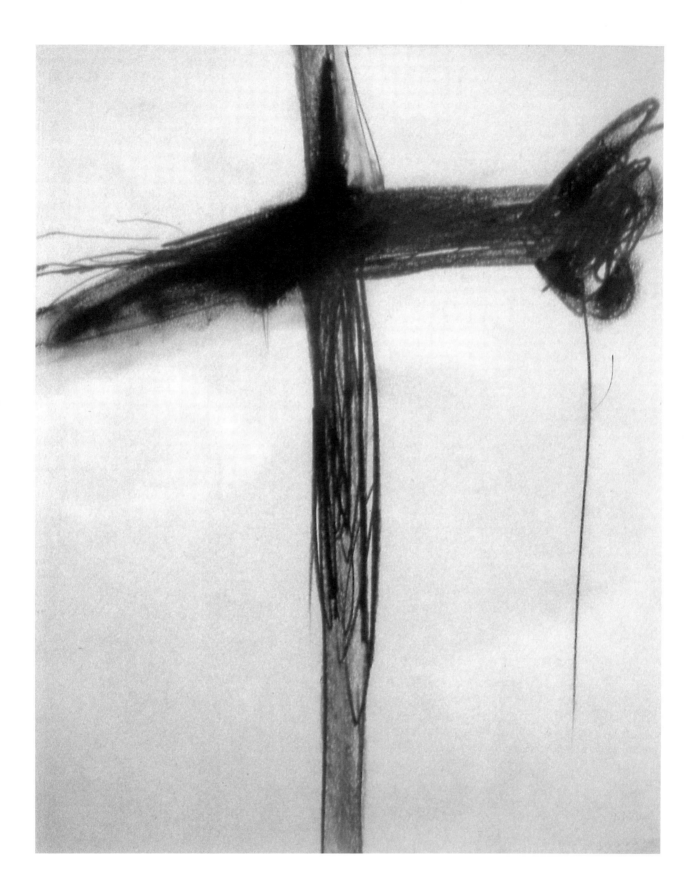

River Sector, drawing by Harriet Brickman. Pencil, oil, chalk on paper.

 # DAY 26

Connections

By studying the recurring harmonious patterns inherent in mathematics, music, and nature, ancient mathematical philosophers recognized that consistent correspondences occur throughout the universe.

—Michael S. Schneider, *A Beginner's Guide to Constructing the Universe, The Mathematical Archetypes of Nature, Art and Science*

We understand the connections between body and earth through experience. Take a moment to feel your breath; close your eyes to heighten awareness of sensations. Breath connects what is out there with what is in here. The outer environment becomes part of the inner environment with every in breath; the inner environment becomes part of the outer environment with every out breath, in a constant process of exchange. Continue to notice the sensations of breathing as you open your eyes. This is inclusive attention—simultaneous awareness of self and the environment.

The book, *Body and Earth,* begins with the concept of wholeness. When we use the word *body,* we refer to all that it means to be human. These characteristics have many names, unique in different languages, for various parts: physical body, mental body, emotional body, intuitive body, spiritual body, soma, psyche, and soul, or collective terms such as person, self, and I. In this work, we consider the word *body* to include the whole. All the parts are essential to live a full life. The same is true of the earth. We use the word *earth* as an inclusive term referring to all aspects of the globe. Again, sometimes the earth's aspects can be described separately as ground, sky, water, plants, and animals; academically as geosphere, hydrosphere, biosphere, and ecosphere; or inclusively as planet, Gaia, globe. In our work, earth includes the whole. Body is part of earth, and earth is an aspect of larger spheres: the solar system, galaxy, and universe.

The second underlying concept of *Body and Earth* is intrinsic intelligence. Rather than considering the body as object, machine, commodity, or property, as merely a resource for our use, we think of it as subject, with its own organic history and integral logic. The same fundamental attitude applies to the earth as well; we consider that it has intrinsic intelligence, subjectivity, and interiority. It is through the body that we experience the earth and through the earth that we understand the body. One aspect may be more accessible than the other; we enter wherever we can. For example, one person may be drawn to the landscape yet feel discomfort attending to body; another is involved with body but distanced from physical context—the earth systems that support life. As we interact with the air, plants, water,

Yoga: Triangle Pose, Trikonasana

CONNECTIONS

Speaking with Mitch Lansky, author of a book about forestry in Maine, called Be-yond the Beauty Strip, I mention my inter-ests in body and earth. He responds, "Body is the environment; it's your first environ-ment." Later, he continues: "The environ-ment, as I see it, includes people. Not just their physical presence but all that comes with their culture. I find it odd that so many people do not see their bodies (or minds) as part of the environment. They therefore do not realize that some of what they put into their bodies is a form of pol-lution. To them, the environment is some-thing else, somewhere else." To make his point, he describes a cartoon: A fellow is driving along a road that becomes rural, with forests and streams. There is a sign that says, "You are now entering the environment."

trees, soil, and animals through breathing, speaking, eating, walking, drink-ing, bathing, touching, and building, we come to know ourselves through the larger context of earth, supported and imbedded.

The third basic concept relates to perception: where and how we place our attention affects what we perceive. Through the senses we construct our view of self and of the world. Perception is filtered by culture, by religious and educational experience, by family and personal history, and by expecta-tions. As we begin to identify and explore our perceptual biases, we recog-nize that the way we focus our senses at any moment affects our experience. We also affect what we perceive. The theory of relativity reminds us that place and people are changed in the moment of our attending.

To understand these concepts at an experiential level, we practice a place scan, engaging senses, memory, and imagination. Begin by bringing your at-tention to water. Where is there water in the immediate environment where you sit? There may be no lake or ocean in view, yet there is water in your body; 70 to 80 percent of the human form is water. Indoors there is water in the air as humidity; there is water in the wood floors and doors, absorbed from the air. There is groundwater under the building, drinking water flow-ing in pipes. Through the window, notice water in the clouds; the nearby pond, stream, lake, or ocean; the plants, animals, and people.

Now focus attention on plants. Where are plants in this environment? In-doors there may be houseplants contributing to the oxygen in the room, as well as herbs, vitamins, and medicines. There may be no living plants, but the floors are made of maple or oak; curtains and clothing are constructed from plant fibers such as cotton or linen. There is plant material being di-gested in your stomach, your shoe soles might be rubber from tropical trees, and your books and papers are processed pulp. Outside, you can see a va-riety of plants, such as trees and grass sustaining life.

Shift your attention to earth and soil. Where is earth in this place? In-doors there may be visible rock—granite or marble from local quarries as tabletops, cutting boards, walls, or fireplaces. There are nails, bolts, and metal fittings of all sorts; melted sand in the glass windows and mirrors. There are minerals in your bones, from the Midwest soil perhaps, calcium and phosphorus transported through the foods you ate for breakfast. Metal fillings or braces may be part of your teeth; gold, silver, and gemstones em-bedded in your jewelry and watch. There is a concrete subfloor to most buildings, and bedrock below supporting the whole. Outdoors you can see the contours of the soil and rock in the landscape, protruding bedrock, gla-cial erratics, sand, or dust. Particles in the sky make air visible, fragmenting light.

Focus your awareness on animals, your relatives on this earth. Where are the animals in this place? Indoors there may be pets, like dogs, cats, or birds; you too are an animal. Animal products may be part of your digestive system at this moment, from recent meals. There may be flies, a hidden spi-der, or mites. If you are in an educational environment, animals might be caged for research, dead for dissection. Outside, if it is spring, there may be frogs and salamanders emerging from hibernation, squirrels busy collecting food, migrating warblers returning from South America, and snow geese headed to the Arctic tundra detailing the skies. Bass could be on their spawning beds in the nearby lakes. If you venture out at dusk or dawn, a plethora of animals might greet your eyes, invisible during the day except for telling footprints.

Return your awareness to breath, and focus on the air connecting inside and outside. Perception is different for each person at every moment. What can you smell? Air conveys chemicals that bond with receptor sites in your nose. You might notice a whiff of perfume or the scent of shampoo from a neighbor or smells you are not conscious of, like butyric acid or phero-mones, stimulating autonomic reflexes. What can you hear? Sound waves travel through air to your ears and vibrate your tissues. Light waves trans-port the energy from the sun or from electric lights to your skin and to the retinas of your eyes, triggering your pituitary gland for metabolic activity; dark corners stir the pineal gland for cycles of sleep. The weather is in mo-tion, connecting one region of the globe to another. Air has pressure, tem-perature, and humidity, stimulating touch receptors in your skin.

Through all of this experience, you are sitting in your chair. Yet you are moving. The earth is rotating on its axis, circling the sun, circumambulating the solar system. Gravity pulls your body toward the earth, centrifugal force spins you out toward the universe, electromagnetic fields penetrate your structure and interact with your nervous system, air pressure keeps you from exploding, and compression stimulates bone growth. This brief place scan can be done anywhere at any time of the day to remind us, at an expe-riential level, of the interconnectedness of body and earth.

We are nature, too, subject to the same physical laws as all other compo-nents of earth. A nerve branches in the same way a tree branches, because

Traveling the east-west trail of Moraine Park in the Rocky Mountains, I follow the meandering path by the stream. A bull elk with enormous rack stands in distant grasses, teaching me about the balance of skull on spine; the strong wings of a red-tailed hawk above detail how arms connect to ribs; the brown trout darting from the shadowy overhang of stream bed displays essential agility of spine.

This is all I need to feel grateful: the an-cient beauty of rock, the humility and pride of standing in the same field as the elk, the vitality of fast-flowing water. Peering at the sandy stream bottom, I see a footprint of the elk and acknowledge connection. It is imperfect, this story; yet as I meander along the path linking me to the family of things, I feel not alone but united. We each know the way, following the current as it flows, joining, whenever we can.

Writing about connections, I hear the news that a student has been in a motorcycle crash. All the complexities of the moment come to a halt, replaced by an expansive sensation. My heart floods with feeling, connecting outward. This edge of life and death accompanies us throughout our jour-neys. The trees whisper with the breeze, water tumbles over the falls, my husband's voice hesitates on the phone downstairs, all merging into a larger sea of sound and si-lence of the unknown in each moment.

Drawing by Caitlin Clarke, age ten.

they are governed by the same forces. Fluids flow the way water flows, because they have the same origin. Body is home. It's our first environment. It is the medium through which we know the earth. Our relationship with our bodies affects our interaction with the world. As we feel seen, we can see others; as we acknowledge depth in ourselves, we can recognize the integrity of other systems. To inhabit place with integrity, we inhabit ourselves cell by cell, recognizing our role in larger systems. Our breath, blood, muscles, and bone are of the earth; they are the air, rivers, animals, and minerals inside us, not separate but same.

✳   ✳   ✳

TO DO

Comings and goings*
30 minutes

• Create a map, tracking the plan of your day. Include major landmarks, places you spend time, pathways that link these places. Indicate sites that make you feel most connected to the area where you live. What stands out in your memory as you consider your days: colors, objects, smells, places, buildings, corridors, doorways, rooms, people? Include these in your map. 15 min.
• Write about your map, and read aloud to yourself or a group. 15 min.

Nature and culture
30 minutes

Choose one topic in each category, and write for three minutes from the perspective of nature, then from the perspective of culture for each choice. There will be eight three-minute writings:
  1. Spider, earthworm, scarab. (choose one)
  2. Crystal, diamond, amethyst
  3. Wolf, coyote, eagle
  4. Cactus, oak tree, palm tree
For example, from the perspective of nature, a spider has eight legs, has silk glands so that it can weave a web to trap other insects, and is food for birds and other animals; from a cultural perspective, the spider symbolizes a wise figure or trickster, spins webs photographed for their beauty, causes fear in many people, and is used on Halloween to signify spooky places.
• Read aloud to yourself or to a small group. Notice both shared cultural assumptions and those that are unique to your family or person.

*Exercise from Ann Kearsley, landscape and urban designer and set designer for dance.

Photograph by Erik Borg.

Place visit: Attention to connections

Visit your place, noticing your familiar pathways and process of arrival. Settle in and begin witnessing place, nonjudgmentally. Notice what you notice. Be as present as you can to the experience of place. If your mind wanders into planning or remembering, return it to the present moment. Engage all your senses, noticing connections. What are your preferences and aversions; do emotional responses surface? Remember, perception is selective; we can't be aware of everything. 20 min. Write about your experience. 10 min.

WRITING HOME (2000)

This is where I lived when my father died, a friend said. This is the place I call home. Landscape fixes the details of memory in space and time. So do muscles, brain, bone. I have lived in New England for twenty-three years. This is where I lived when my father died. This is the place I call home.

My life in New England is a triangle of place: a community in Massachusetts, a college in Vermont, a wilderness in Maine. Native peoples migrate with the seasons. I too am a gatherer: blueberries, apples, seeds.

Does place change who we are, what we think, how we move? How could it not? Cells exchange molecules with every breath. Outside is inside. The root in the garden becomes our teeth.

We are all recent arrivals. I have journeyed to ancestral homelands: an island off Denmark, along the River Rhine in Germany. They were as familiar as a stroke of hand along the curvature of my spine.

So too was the Olduvai Gorge in Kenya. Standing on that dusty soil, I could easily imagine walking hand in hand with my mate; just as I recognized fellow tourists, Japanese, Tanzanian, and French, as family.

How do we recognize home? Where are you from? Do you know? Do we care? Are we there?

Read aloud, or write and read your own story about connections.

DAY 27

Motion and Emotion

We live our lives in a sea of emotions. Sometimes we don't consciously feel them; usually we do. They are always there.
—Joel Davis, *Mapping the Mind: The Secrets of the Human Brain and How It Works*

COMPLETING THE PICTURE

When my high school boyfriend left to study in Japan for a year. I didn't know I cared. The next day, I had chest pains so severe that I was restricted to bed. The doctor prescribed muscle relaxants, treating the constriction around the heart. As I lay there in bed, feeling alone and despondent, I became aware of the emotional component: I was sad. It was the first time I understood that emotion has sensation; heartache is real.

We track emotions in the body through sensation. The screech of a car or the sight of a baby's smile register through our senses, initiating a series of interactions within both brain and body. Although we often consider emotions irrational, neurobiologists tell us that they are intricately interwoven with the circuitry for rational thought. Emotions are relational, essential to humans as social creatures. Because emotions involve the speed and thoroughness of the endocrine system, we often describe them in the language of water, as fluid states: a flood of feelings, a wash of emotions, a flow of warmth, a surge of tears, a wave of happiness; or as a process: bubbling with laughter, drowning in grief, frozen in fear, or swimming in excitement.

Emotions are happening all of the time whether we are consciously aware of them or not. Every emotion has accompanying sensation. As we heighten awareness of sensation, we can "listen in on" our body's internal response to our mental processes. Allowing emotions their expression can feel like riding a wave. Look at something that creates a strong emotion, a photograph or a landscape, and notice the sensations in your body. Let your response ride on a breath into expression: "Ohhh!" If you follow the waves of emotions rather than trying to control them, you can become familiar with the gathering of energy, cresting, surging forth into expression, and the dissolve and return into the oceanic matrix of the body. Each phase is full of sensation, energy, vitality, and risk.

Mood is prolonged emotion, a feeling state that pervades the body as the endocrine secretions move through the system. That is why movement can change our moods; it facilitates the pumping and processing of hormones and stimulates a new set of directives. Mood can be the pre-effect or the afterwash of an experience: a holding back of a wave of emotion or the resulting calm after hysterical laughter, an angry fight or a good cry. Emotions involve our joy and irritation, not just our thoughts about joy or irritation. They are experience; they are not abstractions of experience. Thus, it is difficult to separate emotion from identity. The usual expression is "I am angry," not "I am thinking angry thoughts." Yet attention to sensation helps us move through and with changing states.

Part of finding home is locating a place where our emotions can live. Our

culture and educational systems are particularly disdainful of emotional life, beyond an acceptable mild range of conventional response. Where is awe, amazement, delight; where is sadness, confusion, despair, and all that comes between? Where is there room and space and permission to feel? This may be in the woods; inside your house, apartment, or room; on a stream in a boat; dancing, reading, or dreaming alone or with others.

From an evolutionary perspective, emotions are foundational. Of the three basic characteristics for life—the capacity to reproduce, to metabolize, and to respond to changes in the environment—the last reflects the under-pinnings of emotion—responsiveness. At a basic level, this capacity and mo-tivation of a cell to move toward or away, to ingest or not ingest, is the sur-vival core for more complex emotional life. Pair bonding and social grouping, reflecting sustained emotional affiliation, came later—largely in birds and mammals.

From a physiological perspective, scientists looked for many years for the one structure responsible for all emotions in the human body. Eventually, they understood that it is the networking between various physical struc-tures and our interpretation of the body's response to these processes that delineates our emotional lives.[1] The limbic system, located at the base of the cortex, has been identified as largely responsible for processing emotions, including memory and learning, providing a foundation for all aspects of

When we faced the enormously difficult task of putting my father in a home for Alzheimer's disease patients, my mother be-came very ill with a cough; she could barely speak. As his mental and physical health rapidly deteriorated, Mother began to re-cover. A few months later she had both knees replaced, an operation that had been put off in the three years she cared for him at home. This required that she attend physical therapy three times a week, with a staff encouraging her to develop strength and get back on her feet. By fall, when my father died, she was fully on her feet, able to take up life alone. Emotional and physi-cal change is interwoven; it is useful to ac-knowledge the integrity of our structure. Any aspect can be an inroad to understand-ing and healing the whole.

thought. Often referred to as the "emotional brain," the limbic ring is not one structure but a collection of structures present in animals and particularly developed in mammals.

Of the various components comprising the limbic ring, the *amygdala* is considered key in the memory aspect of emotion. Described as the emotional sentinel of the brain, it screens incoming sensory information for emotional relevance, sometimes overriding higher thought processes. (It also stimulates and soothes tears!) The *hippocampus*, shaped like a sea horse, determines our general "felt sense" by integrating various sensory modalities. It also cements in place (over a period of hours or days) the network of sensations and memories that equal a particular person or event, informing future survival decisions, and helps call up emotional memories. Overall, the limbic system is involved in our "emotional intelligence," including alertness, emotions, passion, tears, sexual feelings, docility, and affection. Thus, it plays a key role in our relationship to place.

To begin to understand emotions, we need to consider three factors: what triggers the emotion, how the brain and body respond, and how those responses then affect the way we internally experience the emotion. Thus, we begin with perception. Emotions occur in relation to a stimulus from the environment—inner or outer, mental or physical—perceived through the sensory system of our body. For example, when you hear a strange sound in the nearby woods, the stimulus enters the ears, is translated into electrical impulses (the language of the nervous system), and travels through auditory nerves to processing centers in the cerebral cortex. Thus, in our example, the sound in the woods is the initial trigger for the emotion.

Then we consider how the brain and body respond. Drawing on Joel Davis's description in *Mapping the Mind,* the auditory stimulus would travel directly to the thalamus, the preliminary relay station for all sensory input, and on to the auditory cortexes in the temporal lobes. After preliminary processing, information goes simultaneously in two different directions, to the frontal lobes (which process and evaluate the information conceptually) and to the amygdala of the limbic system (which screens for emotional relevance at the level of survival, determining the speed and complexity of processing). If the stimulus is sufficiently strong to require immediate emotional processing, it will proceed directly to the amygdala.[2]

Two different neural pathways transmit the information from the amygdala: one to the brain stem (affecting neurotransmitter chemicals that affect the brain's responses) and one to the hypothalamus, which interacts with the autonomic nervous system and the endocrine system. Thus, both brain and body are aroused and involved. Interpretation includes assessing the quality of the sound and associating it with past experience—in this case, experiences and information about sounds in the woods. If it is a familiar sound, like a squirrel rustling in the leaves, a very different set of signals is transmitted at this juncture than if it is an unusual sound, possibly threatening, such as a bear or hunter moving through the undergrowth. In our example the sound is unusually loud, stimulating the release of hormones and chemicals to the brain.

Now we begin responding to our own responses! The body sends signals to the brain about its state of alertness, such as changes in breath, tightening of stomach muscles, and dilation of eyes in preparation for threat. The signals reporting the response of the body and the conceptual interpretation of the brain (frontal lobe and amygdala) merge to create what we experience, and name, as an emotion. As Joel Davis says, "The collection of alterations in the state of the mind and body which these neural circuits detect is the essence of an emotion. The feelings that result—fear in this case—is your experience of these changes."[3]

Identifying and naming emotions is a learned activity. The connection between the multiple sensations registering in the body and conscious thought, including language, develops after birth. When a baby is born, emotional aliveness is clearly visible. The expressions of face and body, including the wriggles, stretches, and explorations of fingers, tongue, and toes, are accompanied by vocalizations from gurgling to screaming. The dialogue between visceral functions and muscular expression is already built in. (In fact, motor nerves are the first to myelinate in the developing embryo. We move to feel.) Yet neurobiologists tell us that the brain is insufficiently developed at this phase for there to be thought connected to this emotional panoply.

Between birth and twelve to fifteen months, as language becomes part of the expressive communicative palette of the infant, the brain grows neurological pathways between emotional centers and the neocortex to link feeling, thought, and expression.[4] In other words, we learn at a very young age the sensations we associate with a felt state of experience, and we then name one state of sensations and responses as fear, another composite as happiness, and so on. As connections become codified in our neuronal networks between thinking and felt states, our basic emotional matrix is established. Because the brain continues to grow, change, and respond, making and losing connections throughout our lives, we can learn new pathways at any point, building on or rehabilitating those firmly established in the first year of life.

Thus, we might say that our mental interpretation of felt states is called an emotion. Our response to an emotion is based on cultural, familial, and personal experience and our expectations at the moment, and it actually creates our feelings. Returning to our example, you might see a black bear emerge from the woods and experience fear as the basic emotion. But once you see that the bear is intent on its own activity and remember that black bears are rarely dangerous to humans, the feelings that result might become a blend of caution and curiosity, humor and delight, inquisitiveness and prudence in response to the experience of witnessing a bear in its own habitat. Although certain emotions are basic to survival, our attitudes about those emotions can and do change throughout our lives, affected by our maturation and life experience. For example, if you have visited a culture where the bear is a totem figure, a sacred animal, you might even consider the bear as a spirit guide, engaging spiritual dimensions. This change in information affects how you internally experience the emotion.

When leading an Authentic Movement workshop, part of my work is to witness movement with a nonjudgmental mind. In one session a woman dances around, flitting through the space, playing and enjoying herself enormously. I feel myself growing irritated, Why can't she be more serious, work harder? I think. But my role as witness also requires that I track these inner responses as they occur. Watching the woman, I recognize how critical I am of the part of myself that likes to celebrate. Things are fine as long as I am working hard! Noticing this, I soften my gaze, enjoying the "me" in her that is free to play, spin, dance like a child.

Emotions are relational in model, not isolated. Various psychological models have been created, based on animal and human studies, to identify "the primary emotions." The generally agreed upon spectrum includes love, hate, anger, fear, and pleasure. Named here in English, emotions are described in diverse ways in various languages. Yet this narrow spectrum hardly represents the richness and complexity of human life experience and does not account for the role of motion in emotion as feelings blend, change, transmute, respond, and interact. A much richer vocabulary is needed, poetic and evocative, representing process rather than implying a static state, which is alien to biological life.

Naming emotions can be useful or limiting. It is often a relief to name a felt state in the body so that so you can be more conscious yourself and communicate more clearly with those around you. To say, "I feel grief" when you stand amid a clearcut forest or see an animal hit by a car or read about a trauma in the newspaper gives you resources for response. This may include a book detailing the process of grieving, stories by others with similar experience, a contemplative practice technique for staying with your feelings long enough to let them move through your system, or a creative form for expression. Most of us have developed our own modes for recovery, however subtle—gardening, reading, running, and swimming can be resources that help us move through difficult emotions. But labeling can also limit experience. When does fear become curiosity; what is the edge between happiness and bliss; where is the distinguishing line between anger and uncertainty?

Our tendency in American culture is to pile all emotions into a few categories, with a hierarchy of importance. Sometimes we use prescription drugs to stabilize their characteristics.[5] So many aspects of our human potential are homogenized or denied existence. For example, soft drink ads show young happy folks with drinks in hand enjoying a carefree life; college catalogs show smiling students amicably cavorting. Where are dreaming and imagining, confronting and questioning, storming and challenging, crying and comforting, subtle reflection and contemplation, creative questioning and curiosity, and the investigation and isolation necessary at points in our lives for deeper understanding of the larger cycles of life and death? Instead, if we really attend to felt states in the body, we can embrace change and emotional fluctuation essential to the growth that is already occurring within us. We can acknowledge that all emotions are part of life, representing a continuum, not a polarization. As we allow the panoply of possibilities, we recognize felt states that are complex, rich, and full of movement.

From a creative perspective, emotions are doorways. If you feel resistance, anger, dullness, you may be experiencing what is called a gateway emotion. This is a strong emotional state, such as anger or irritability, that takes hold and permeates the body. When entered, it leads to other, more diverse emotions and sensations. Be fierce; acknowledge your resistance and move through it. Also be gentle; we have developed gateway emotions for good reasons. It is important to recognize their use in our lives, protecting us from feelings that are overwhelming, dangerous, or beyond our integrative

Door, architectural detail by Anya Brickman Raredon, age sixteen.

capabilities. As we open doorways, we increase our resources. When we can surrender to a feeling of participation in our emotions, creativity abounds.

A healthy emotional life requires movement and expression, especially when the powerful endocrine system is involved. "Shaking it off" can be an essential component of health after trauma. Think of all the times in your life when you have had experiences that felt traumatic—even falling down on a sidewalk, falling off your bike, giving the wrong answer in school, or observing a family fight. Immediately you were expected to look and act as if nothing had happened. All that impact gets stored in the body. If you instead take time, shake it off, allow twitching or trembling, the experience is released from your systems, and you can move on. Watch your dog after a scolding, an athlete after a fall; they generally shake it off and resume their activities.

The inhibition mechanisms for emotion in the body are both useful and dangerous. If we reacted to everything that stimulates our senses, we'd be an emotional mess. But if we repress all that we feel, this too is exhausting and dangerous to our health, our relationships, and our community. It takes a lot of energy to hold back emotional expression, often creating a stagnant pool of irritability or unnamable darkness. Unexpressed feelings will find expression in indirect ways. For example, you yell at your dog because you can't yell at your boss. We need to find places for expression, release, discharge, and recovery in our busy, emotion-rich lives that are appropriate, rather than thrusting them unconsciously on others or the environment.

Emotions play a key role in our relationships with body and earth; they are inextricably linked with rational thought and moral choice making. We must be cautious, in this fast-paced world, that we don't dampen emotional responsiveness and lose our sense of responsibility. When we cease feeling, we stop caring. Heightening awareness of sensations—the language of emotions—helps us to feel our responses to each moment. As we can feel sensations, we can make informed choices, allowing our bodies to guide us responsibly in our interactions with the world.

✳ ✳ ✳

Tracking emotions
15 minutes

TO DO

Seated, eyes open or closed:
• Focus on breath and notice emotions. For example, "Oh, this is pleasure." Notice the sensations of pleasure in your body but stay focused on breath. You may feel tingling or warmth particular to the moment. "Oh, this is anger." Notice how anger feels in your body: there may be heat, tension, or big volumes of energy. Greet your anger. But keep your attention focused on your breath. "Oh, this is fear." Notice how fear feels: there may be nausea or constriction, and your heart may speed up, or breath may be shallow, palms sweaty.
• As you observe emotions and sensations without reacting, notice if they change or blend.

- Begin naming emotions. Name as many as you can; include those you are feeling in this moment and those you remember from other situations. Say them out loud and feel their presence in your body. Spend time with the feeling states that accompany emotions, such as acceptance, anger, anticipation, disgust, fear, grief, happiness, joy, sadness, surprise, alertness.
- Now take three minutes and really talk yourself into a bad mood. Then take three minutes and talk yourself into a good mood; shift your state of mind. Notice your capacity to change mood with mind.
- Assume a posture that makes you feel depressed. Find the body state for bad mood. Then shift into movement that puts you in a good mood. Feel the transition. Remember, movement affects mind, and mind affects movement.
- Write or talk about your experience. 5 min. How does expression affect mood?

WRITING HOME: THE STREAM

Before heading west, we visit the stream near our house in Maine. We have come here today to exchange wedding vows once again, merging two lives into one year after year. Sitting on this rocky bank, we watch pristine waters tumble over Little Falls and wait for words. Rivers are connectors, one person to another, one era to the next.

This path where we sit was once an ancient Indian carry. Men lifted canoes onto their shoulders as families walked one by one along this three-mile stretch of rocky water connecting lake to lake. Ritual objects and artifacts have been found that tell of the Red Paint People and those, the Passamaquoddy, who followed.

I am turning fifty; will travel to the Grand Canyon and the Colorado River to mark this event. At dawn, I will drive west over the Piscataquis River and Pushaw Stream, then fly east over the Kennebec to Boston, west again over the great muddy Mississippi of my childhood, before arriving in Las Vegas, all lights, and descending to the canyon.

Shaking it off
10 minutes

- Start with the top of your head, shake it lightly.
- Carry this movement down through your skull, jaw and neck, shaking off any excess energy, tension, strain. Breathe deeply and fully.
- Spend time with your shoulder girdle, gently shaking all the way down your arms and hands.
- Include your ribs, lungs, heart, and diaphragm.
- Continue down your spine and pelvis. Include your belly and all its organs. Release the muscles as you shake, including the gluteal muscles (your butt) and pelvic floor.
- Shake the thighs and knees until you feel the thigh muscles start to release on the bone.
- Add lower legs and feet, shaking one at a time to release all tension.
- Jiggle the whole body.
- Slowly bring all movement to a close, and notice the flow of sensations through your body; notice the emotions that accompany these sensations.

Every hour during the day, find a private place where you can release stored emotional movements. Use sound and movement; notice what comes. Sometimes you don't even recognize what you've stored in your body—it may be someone else's emotion! As in cleaning out your closet or deleting unnecessary parts of a sentence, you need to take time to clear your emotional palette. How do you do this in your day?

See also Peter Levine's *Waking the Tiger: Healing Trauma, the Innate Capacity to Transform Overwhelming Experiences.*

Colorado Sand Dunes. Photograph by Josh Keith.

Place visit: Attention to emotions (riding the wave)*

As you walk to your place, notice emotions. When you arrive, take time to clear your emotional palette. Shake off the emotional residue of your day. Then allow a fresh emotional response to your place, its colors, smells, textures, and rhythms. Emotions may be pleasant or unpleasant: there may be a new blossom or a familiar animal; there may also be litter at your place, the sound of trucks nearby, or a dog barking incessantly. Identify one emotion and notice its accompanying sensations in your body. Breathe deeply.

Continue allowing the breath to support your emotion. Begin a spinal undulation, small or large. Notice what happens to emotions when they are allowed to breathe, move, flow through the body. If riding the wave feels like too much, pause in open attention. You are creating a dialogue with emotion; you don't want to be overwhelmed. Acknowledge your choice: ride the wave, rather than plunging in over your head. Notice how emotions inform your relationship to place today. 20 min. Write about your experience. 10 min.

*Exercise developed from Susan Harper's work, "Em'Oceans and Sensations."

On the third day, I will kneel in the inner gorge amid cold currents, pressing a black stone two billion years old in the palm of my hand: Vishnu schist. Our young river drivers will spin our raft 360 degrees, our minds scrambling to grasp a time so vast, as our eyes trace the rim one mile above. And on the last day, we will float on the manmade lake, see the giant dam—the works of our time, our people, and fifty years will seem a very short time in the picture of things.

Yet today, we stare at bubbles rising from swirling pools amid glacial boulders. Already anticipating the spectacular, I say to my husband and also to the river, this year I will listen more carefully.

Read aloud, or write and speak your own story about emotions.

DAY 28

Sensuality and Sexuality

There is . . . a common sensibility shared by persons who have, in Robinson Jeffers's phrase, "fallen in love outward" with the world around them. As their compassion for the land deepens, they . . . rejuvenate their senses by entering into reciprocity with the sensuous surroundings.

—David Abram, *The Spell of the Sensuous*

LONGING

My friend returns to the Grand Canyon again and again, called by its presence. Her desire is to merge with the landscape, to enter its crevices, submerge in its waters, experience the awe of union and separation with each visit.

We connect to the environment with rich sensorial experiences, layered with smell, sound, taste, visual stimulation, and emotional arousal. Through all of these experiences and more, we build our relationship to place. Embedded in the environment since birth, we may sometimes not remember what has caused our deeply felt response to place. Yet our bodies hold memories, and certain landscapes draw us to them in a mysterious pull, calling for union. This romance of place, experienced breath by breath, greets us each day, teasing our conscious mind, framing decisions, and shaping our lives.

Place is palpable. We live in a specific context, feeling all that is around us. Perception interweaves past with present and with expectations about the future. The warm sun, pungent smells, and caressing wind are felt each day, even if our focus is elsewhere. Yet at any moment we can direct our attention to the possibility of pleasure, allowing a sensuous relationship to body and earth. We invite the eros of place to penetrate our awareness, bringing our attention to the exhilaration of the moment: the voices of the finches tittering in their cage, the plush cotton of a bathrobe, or the deep breath of a dog curled by our side. Acknowledging the erotics of place means experiencing life through the lens of that which stirs, arouses, and pleasures the body.

Sensuality is free, bereft of economic value. Like clean air, pure water, wild plants and animals, what is most valuable, most luxurious is often overlooked. Our culture breeds dissatisfaction in order to sell products. Our image of sensuality and sexuality is often associated with arousal by something outside our basic nature, couched in neediness or competition. Yet we don't have to fly somewhere else for spectacle of place. We don't need special perfumes, exotic cars, or a new affair to stimulate the senses or affirm our worth. Sensuality is a process, not a thing; an experience, not a commodity. What we need is to bring the expectation of adventure, joy, and pleasure to perception of the moment at hand. As we realize that we are sensual beings, living as part of a world that delights the senses, we can distance ourselves from the cultural baggage of dissatisfaction. We need less, not more, to experience a full life.

Torso, sculpture by Herb Ferris. Schist and chestnut.

    While sensuality enhances our relationship to self and other, sexuality engages our capacity to reproduce our species. Drawn by lunar rhythms and affected by changes in light and dark, the hormonal secretions motivating reproductive desire are powerful forces in our lives. Sexual energy is available to each of us to employ as we will. Energy itself is of one root, part of our biological makeup. The same energy can build a house, save a rainforest, write a book; or it can start a war, kill a neighbor, rage at a child. Cultures throughout the world have devised ways of shaping, diverting, controlling, and interacting with this most vital source for the good of humankind and the health of the planet.

    What is not useful is to remain naive about our responsibility in conceiving

In a voice workshop with my writing group of six women, we are asked to speak, beginning with the words, "What I find sexy about myself is. . . ." and fill in the blank. After a few minutes of jokes and humor, we begin. When it is my turn, I feel energy rising in my body, up to my throat. I don't think I can get a word out of my mouth. Looking at each person in the group, I finally say "silence." "What I find sexy about myself is my silence." The leader responds, "No wonder you became a dancer!" Within the group, the statements are so unlike the cultural norm of sexiness, yet a radiant beauty emanates. Sexiness, it seems, is connected to speaking truth.

a child. In this time of overpopulation, with humankind doubling in numbers every thirty years, every woman has to choose whether it is wise to give birth to a child. Every man has a choice about whether to create a child. Although sex is perhaps our most creative act, ultimately it requires attention and decision. It is one thing to unite sperm and egg; it is another to sustain the lifelong process of child rearing and parenting. Humans have long gestation periods and long parenting cycles, with unprecedented economic implications. Ignorance of body and ignorance of earth promote fear, which itself encourages abandonment of power to choose. We cannot remain ignorant when our actions have such far-reaching and immediate impact on others and the planet.

Sensuality also involves risk; feeling and caring heighten the possibility of being hurt. If you have opened yourself to love and felt betrayed, you may not want to experience that level of pain again. If you have cared deeply for a place, bonded to a home, a field, or a favorite tree, only to move at age seven or see your land sold, you may not easily invest your feelings in another place. Pain from loss of home and habitat is more pervasive than we might think, a part of our individual fabric that we might not remember or recognize.

Dissociation is separation, distancing from bodily senses. Dissociation is often described as a kind of hovering above, watching rather than embodying the physicality of the moment. Sourced in a traumatic experience, emotional or physical, that has violated our sensual and sexual selves, dissociation begins as a form of protection. If one is absent from the body, the violation is less personal; dissociation provides a place of hiding when none is present. Prolonged dissociation is sometimes accompanied by violence to self: cutting, scarring, making the unspeakable pain visible. Recovery from dissociation is a slow and critical process. It requires a descent into feeling, a reinhabitation of the body, a return. Dissociation occurs in relation to place as well, distancing ourselves from a landscape or home, perhaps by moving frequently or remaining an onlooker to community and connection. Sometimes we destroy a landscape, making our distress visible. Recovery involves a careful process, reinhabiting a sensual relationship to place while honoring the pain.

In some ways, all of us are dissociated from our bodies and from our places. It takes constant attention, working against the flow of culture, the pressure of time, and the context of productivity to engage our sensual and sexual relationship to self, other, and place. What was once a natural heritage now requires nurture and support, perseverance and resolve. As we reflect on deep attachments at different times in our lives, we can begin to recognize the degree to which place—or the loss of place—underlies our present responses and feelings. Sometimes it is not until we deal with the grief of losing a parent or child, of losing the corner store or the childhood farm, that we can engage deep feelings toward people and place once again. Acknowledgment helps free perpetual numbness, unfulfilled longing, and repression of spontaneity. Sometimes we can write ourselves home, listing and listening to experiences of place to make conscious all that motivates and informs our lives. One role of sensuality is to undo the knots of tension from our lives so that we can open to connection and responsiveness. One role of sexuality is to reestablish our fundamental creativity and connection to forces larger than the personal.

In evolutionary history, cellular division was the reproductive norm for the first few billion years of life. As prokaryotes and eukaryotes evolved 3.4 billion years ago and 1.2 billion years ago, respectively, they utilized asexual reproduction with genetic material duplicated identically in each new cell. The first sexual reproduction occurred around 1 billion years ago, with genes of two parents contributing hereditary material. This change increased possibility for diversity through unusual combinations of genes, and both plant and animal differentiation proliferated. Reproduction is pivotal for survival of species, and reflexive and behavioral patterns are encoded to ensure reproductive success. As mammalian and avian species developed relatively efficient metabolisms, more energy was available for communicative interaction. Group living enhanced survival but also demanded reproductive strategies for the collective as well as the individual. Long gestation periods and extended mothering provided a longer period of brain development, basic to the evolution of higher intelligence.

While sensuality and sexuality arouse and stimulate, they also calm and prolong communicative interaction. A small but powerful peptide hormone called oxytocin has been shown to play a predominant role in bonding. Prior to childbirth, oxytocin stimulates the flow of milk from the mother's breast and related changes in body and behavior. It is the hormone that creates the peaceful union between mother and nursing infant and sustains years of child rearing. It also has been shown to be present in other types of animal bonding, including "parental, fraternal, sexual, and even the capacity to soothe one's self."[1] By extension, it may offer a biological model for our strong feelings toward place. Oxytocin originates in the hypothalamus, part of the limbic ring that also governs sex drive and emotions. As we have seen, the fight, flight, freeze, or friendly response of the autonomic nervous system prepares our defense mechanisms for survival, readying responsiveness by increasing heart rate, dilating the pupils, and suppressing digestion. Oxytocin, on the other hand, stimulates our expansive nature, calming and nurturing our relationship to people and to place.[2]

Sexuality is inherent, but it also is informed by our childhoods and our ongoing lives. Because responses to erotic stimulation are based in infant experiences of tactile stimulation, they connect to powerful emotional needs, which may or may not have been met when we were children. In most cases, sensual delight in the body is alive and well at birth and gradually becomes socialized within what we hope are healthy restraints. (At various periods in our lives, we confront, question, and even break these same boundaries in the process of defining self as separate from, as well as related to, family and culture.) Thus, sexuality remains charged with issues of childhood as well as with an emerging sense of self and other as adults. Feelings about dependence and independence, abandonment and the need for control, are a few of the possible concerns that affect sexual behavior.

"I'm afraid someone near me is going to die." This line from a friend's poem provokes forbidden thoughts. With each step of deepening intimacy in marriage—like seeing the shape of my husband's head leaning on the oak rocking chair, relaxed; like depending on the warmth of his body in bed; like needing to call each night— any of these patterns toward merging is balanced somewhere in my brain by knowing the pain of loss. Where does this measuring come from? Perhaps it is a survival response, like not standing too near fire, like not owning a dog for fear it will get hit by a car, like not trusting my body to bear a child. Perhaps it has something to do with standing alone, with taking care of myself early on, merging with art instead of people.

Yet my distance from him fades each day we walk down the rocky path in Maine. Past the rotting moose bones, up to the side road where the dog's paw was caught in the fox trap, around the bend where the grouse hang out—each repeated sojourn makes an inroad in my body that is more and more difficult to ignore. Then I must acknowledge that I may want to abandon him when difficulty arises, that I may not be a good caretaker, that I still feel the sullen disappointment at how I met my father's end, my betrayal. Will I do it again if it is my turn to walk in the hard territory close to death?

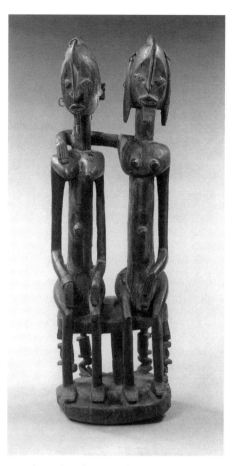

Seated couple. African, Mali, Dogon
nineteenth–twentieth centuries. The Metropolitan
Museum of Art, gift of Lester Wunderman, 1977.
(1977-394.15).

Respect for our bodies—our physical, emotional, intellectual, intuitive, and spiritual selves—underlies sexual responsiveness. The more comfort and trust we have in our bodies, the more choice we have about our sexual actions and interactions. The interoceptors in particular register safety or distress; noticing internal physical signals in various situations informs safety and allows connection. As we explore the fundamental territory of moving away or toward, the yes or no of opening and closing, we can establish healthy boundaries of communication and expression. Respect for others and for place is based largely on this respect for self. Perhaps nowhere more than in intimate communication with another do we need to know that our bodies are with us, supporting our needs and rejecting what is not appropriate.[3]

The modulation of flow between being closed and being open is the pulsing life of healthy sexuality. There is no reason to be open all of the time to people and to place, or to remain closed. The body receives subtle information from its environment and may know more than we can consciously understand about what is safe and what is threatening. As we trust the wisdom of the body to choose when it is appropriate to open, we develop a different understanding of experiences such as frigidity or impotence as well as of orgasm. Adaptability gives us our responsiveness and allows us to trust that our bodies will and can protect us if we listen to the signals. If we are taught to be afraid of our bodies, we lose a vital connection to our inherent knowing. Sensual awareness offers perspective and allows the body to feel familiar rather than foreign. As we follow our curiosity and open to sensations, we build a foundation for knowing, directly, the world around us.

Because of its connection to the endocrine system, sexuality is highly interactive with our emotional lives. Sexual experience can include the many subtle aspects of sexual exchange that we experience each day, based on energies awakened in ourselves as sexual beings. For example, the warmth felt seeing a friend laugh, the feelings stirred hearing a poem read aloud, or the empathy felt watching an athlete or memorable performance can involve our sexual energies. Some describe the climax of sexual union as a small death of self, of boundaries, of who we think we are. As we embrace our potential as sexual beings, we can begin to recognize the wide range of possibilities for stimulation and response that we encounter every day. Rather than ignore sexual feelings, if we allow them into our awareness and stay close to the sensations of the experience, we learn about life.

There are many ways to fulfill sexuality. If we move sex out of the media context of seduction and fame, out of the economic model of ownership and power, out of the puritanical context of sin and indulgence, we may return to our native sensual, sexual selves. From full sensuality, we can love the earth. We can value stimulation that heightens relationship, rather than activities that destroy or disturb, putting others—human or other than human—at risk. Daily we make choices about actions and intentions; we need our sensual selves present to heighten connection to ourselves and to place.

✳   ✳   ✳

Romance of place

30 minutes

Consider a place that feels special to you, that draws you to it and evokes your fullest, most alert self. Consider its location on a specific continent, on this globe.

• Write about your romance of place, using language that is experienced in your body and evoked by the place. Include all your senses and tell how they interact with its features. Keep your vocabulary grounded in what is real, using specific names for plants and animals.

• Evoke a sensual relationship to place as you write. 20 min. Read aloud to this place. 10 min.

Merging with the landscape: Inhabitation

30 minutes

Take a walk and find a place that invites your participation. Explore how you can become part of the landscape so that someone walking by might not even know you are there. For instance, enter a pile of leaves, lie between crevices of rock, or wrap your body inside a shallow cave or around a tree branch. Be safe in your choice so that you can enjoy the sensations. 10 min.

• Disappear into this place, becoming part of its skin and texture. See with its eyes, breathe with its breath. 10 min.

• Separate yourself once again and pause in open attention. Draw or write about your experience. Consider whether release of identity expands awareness of self and other. 10 min.

Place visit: Attention to sensuality and sexuality

Take time to enjoy the walk to your place. When you arrive, enjoy all the smells at your place; note how they register in your body. Enjoy all the textures at your place; note how you experience them. Enjoy the sounds and sights; note how they affect you. Enjoy the inner sensations in relation to your outer experience of place: note how you perceive this dialogue. Enjoy breath, your whole sensitive, sentient body, and being in and with and of this place. 20 min. Write about your experience. 10 min. How often do we take time to enjoy what is occurring, inside and outside?

WRITING HOME: THE LAKE

Our lake in Maine is so old and deep, you can only guess. Ringed by hills, overglazed by clouds, the water below reflects water above. Each day I dip my whole body into cold currents, skin touched in places rarely graced by light. Submerged, I am reminded of all that came before: the swirling of gases, the colliding of planetesimals, the differentiating of layers, the erupting of volcanoes, the raining down of waters, the emerging of life forms, the scouring of glaciers, and the migrating of peoples to this place by the lake that I and others now call home.

At dusk we launch our canoe upon the surface. Wanting to know more, we dip our paddles beneath reflective film and stroke. Traveling to distant islands, the rhythm of breath meets rhythm of boat as we surge and dip over waves. Patterns large and small accompany each moment. The swarming of insects and the fullness of the moon both encourage fish to rise from their depths and to bite.

Read aloud, or write and read your own story about sensuality. Breathe fully, and allow your eyes to take in the people and place around you.

Art and the Environment

First, we must understand that the universe is a communion of subjects, not a collection of objects. . . . While every artist must discover this presence and accomplish this connection in some manner, the best artists are those who carry us deepest into this experience of communion.

—Thomas Berry, "Art in the Ecozoic Era," *Art and Ecology*

DIMENSIONS

The forest around our house in Maine is being cut. Alert, we are confronted by the complexity of human nature. If we get too emotionally involved, we'll just get depressed. Stories that are full of despair don't go anywhere unless they encourage positive action. A Passamaquoddy elder watches the logging trucks pass, loaded with hemlock trees. "They are taking the babies now," he says. At ninety, he still makes birch bark baskets honoring the woods. Many around the globe employ art as a way of caring without shutting down. I pick up my pen.

The connection between arts and the environment is threefold. First, art is about wholeness. It connects disparate parts into a unified view. Second, art explores the unknown. The creative process inhabits the edge between what is known and what is not known, a place of heightened possibility. Like an ecotone, the dynamic edge zone of ecosystems that offers the richness of two overlapping habitats, creativity is a place of great potential. And third, art sensitizes us; as we feel, we also care.

The creative process begins with an attitude of play. To be receptive to creative insight, we relinquish our desire to achieve. We can invite the muse and allow connections, but we can't force or control the process. An attitude of play encourages attention to the moment, rather than outcome or results.

Sculpture by Michael Singer. Concourse C, Denver International Airport. Photograph by David Stansbury.

It evokes a willingness to open our carefully developed self-image to divergent possibilities, to be naked in the sensuous stream of experience. As we look for the moment of delight in the routine, the experience of awe in the familiar, we rekindle connection to the creative forces around us.

Art making requires an act of surrender as well as the rigors of ordering. We are re-created as we create. In the process, we draw on all the experiences we have known—experiences of family, of birth and death, of intimacy and of place—and also from experiences we have never known, tapping our collective imagination. As we risk opening new doorways and facing new perspectives, creative investigation can seem frightening and overwhelming. Even when the process becomes familiar, beginning a new work can feel like entering a vast darkness from which we will never emerge. The creative process is dynamic. As we become familiar with the actual sensations of not knowing, both tactile and emotional, we open to new experiences. Replacing fear with curiosity, we practice possibilities for expression that penetrate our identity.

Although it begins with playful curiosity, art making requires practiced skill in a particular medium. Each creative impulse takes a specific shape, and each shape speaks its own content. Days, months, and years are spent developing the necessary reflexes, neuromuscular patterns, and refinement of perceptual modes that underlie the spontaneous emergence of a phrase, gesture, or motif. Familiarity with the physicality of materials heightens expression through a particular medium, affecting the way pen moves on paper, feet spring from floor, and sound vibrates in chambers. Investigation of both person and materials melds into product. Although a creative life can be described as egotistical or self-involved because of dedicated focus, it also involves merging with others. In many ways, the more personal the investigation, the more universal the findings; the more embodied the experience, the more transcendent the product.

Just as the creative process requires inner attention, it also commands a report to the larger community. If it were enough simply to journey and discover, the poem would be left in the journal, the dance would never be completed, and the symphony would remain in the imagination of the composer. Communication is part of the creative process, even if the product is shared with only one other person. At some point the work of art has its own identity, and the role of the artist becomes that of witness to the wild and mysterious emergence. Part of creativity is returning again and again to the emptiness. Many artists feel they are recipients of or benevolent hosts to the visions, images, words, or songs that move through them.

Each art form has its own relationship to environmental concerns. Through the medium of creative writing, poet Mary Oliver and nature writer Barry Lopez rehabilitate our relationship to both body and earth with stories of the land. Musician John Cage reminds us that silence equals sound, and Paul Winter includes the voices of whales and wolves and the acoustics of canyons in his recordings, enlarging our view of community. Visual artist Andy Goldsworthy arranges stones in unusual settings; Magdalena Abakanowicz includes the human figure as part of the landscape.

A former student, working as an editor, asks me to write an article for an environmental journal. When I send her recent writings, she calls back excited. "Yes, just localize and personalize your views," she says. The irony strikes on various levels: the pleasure of being taught by my student, the ease with which she assumes that stories of place belong in the fabric of academic views, and also the insight embedded in her youthful words, "Be more direct and more personal," offering a challenge to each of us for a lifetime.

A Tibetan student, raised in Dharmsala, India, began her senior thesis, "Youth in the Tibetan Diaspora." In the middle of the semester she stamped her foot in my introduction to dance class and exclaimed, "Whenever I dance, I feel so angry—especially when I show my own choreography."

In creative work, anger is often a doorway emotion. So I suggested, "Consider all you have been through, all your country has suffered, all the topics your thesis immerses you in each day, which need integration. Anger may be the guardian at the gate of your own impassioned voice."

There are many gateways to creative projects—excitement, boredom, irritation, hunger, avoidance, depression, or hyperactivity. Each may signal charged experience waiting for attention. Personal projects help us learn our own creative inroads and process, while demanding impeccable work. But don't stop at the doorway, walk through and see what is actually there.

Dancers Eiko and Koma perform in rivers; Ron Brown, in urban streets; Liz Lerman, in shipyards—creating a dialogue with place. Filmmaker Ken Burns focuses on history and stories. These are artists committed to making the unseen visible, to awakening our senses to vital connections with the earth.

Although artwork generally stands for itself, artists are often eloquent spokespersons for the earth. Those generally classified as environmental artists care deeply enough to engage in the lifelong process of articulating their experience with the natural world, including humans as well as living and other-than-living entities. They involve body and place as essential components of expression. As we are willing to embrace and merge with the environments around us with all of our senses open, we inhabit the land we live in and fulfill our animal nature. As each person communicates her or his findings back to the community, we amplify creative potential.

In our commodified contemporary lives we often act as though art, including the images and creative activity around us, has no "real" impact. Yet images touch the part of the body that speaks in symbol, as dreams do. Carl Jung, in *Man and His Symbols,* writes: "Man today is painfully aware of the fact that neither his great religions nor his various philosophies seem to provide him with those powerful animating ideas that would give him the security he needs in face of the present condition of the world. . . . Since nobody seems to know what to do, it might be worth while for each of us to ask himself whether by any chance his or her unconscious may know something that will help us." As we acknowledge the resource of art in our lives, we recognize its role in community and our potential involvement both as creator and as participant.

In many ways, the core of community is creativity, instilling the sense that each individual is participating in the unfolding of possibilities. Moment by moment, one is responsive to and responsible for the overall effect. Just as the presence of creativity affects community, its absence too has an impact on our daily lives. When commercial enterprises and competition alone form the heart of a society, interactions become empty and lifeless. Human engagement requires more than buying and selling to inspire our humanity; we need cultural models that pose questions, inspire celebration, and encourage participation and integration.

Until recently, art in some way reflected the splendor inherent in the natural world. Works were created by individuals who themselves had access to nature in their daily lives. A tendency toward beauty is prevalent in the environment. Patterns, in both macroscopic and microscopic perspectives, reflect mathematical order, seeming fractals of a cohesive whole. Water, seashells, muscle fibers, air, and even the galaxies themselves are formed of consistent spiral structures. From the elegant symmetry of a chambered nautilus to the slowly rotating Milky Way galaxy, the chaos of human experience is embedded in an implicate order.

Even if the artwork itself challenges or offends, the artist's process includes a weave of inner and outer experience leading to wholeness. Etchings by Kaethe Kollwitz depict the horrors of World War II while instilling faith in the beauty of human creativity. Igor Stravinsky's symphonies were shocking at the time, yet they reflect an inner order. Art is not imitation; it creates a work with integrity, and life shines through on its own. Eventually, the artwork stands alone, without its creator. It has its own presence, place, and breath. For each artist, creativity is about integration, both reflecting and participating in the underlying rhythms of cohesive form.

From an anthropological perspective, one role of art has been that of mediator between the ordinary and the extraordinary forces of the universe. Certainly, poetry and dance, removed from economic exigencies because they rarely make money, have long been seen as a direct link between the mundane and the spiritual. The original call of art was to tell the truth, to inspire truth in others, and to sway the resident gods and goddesses for good crops, fertile soils, and healthful life. This role provided connection to natural forces, an invitation to encounter honesty, and a mode of participation in wisdom beyond human limitations.

But our attitudes about art are shifting. In this century, many Americans view art as a leisure activity or an economic ploy rather than as having intrinsic value. As we commodify art and creativity, we see art as other and creativity as foreign, rather than familiar. In the process, we risk losing access to our deepest visions and our intuitive resources. Conversely, as we reclaim creativity, we engage the unknown on a daily basis and listen for truth. In this time of environmental crisis, we seek a new relationship to nature, to our bodies, and to our creative voices.

In human evolution, through our bipedal, tool-making, language-using way of life, human brains emerged with the capacity to create music and art, to deduce mathematics and to reason. Yet this same creative potential has also produced destruction of the earth. The two branches grow from the same trunk. In the past century, when war was the norm on our planet, journeys into the unknown reflected our own destructive powers. While commercial media insist on portraying the constant smile to sell products, art makes the shadow visible. In the past, stories or paintings included favorite landscapes remembered. Now they are eulogies to a meadow destroyed, the giant sequoia cut down, or the manatee threatened.

Retained from our reptilian ancestors, the limbic system of the brain registers newness. One structure in the limbic ring in particular, the amygdala, reacts to anything out of the ordinary, unusual, or different. Limbic arousal alerts the autonomic nervous system for survival responses in case of threat, stimulating a readiness in humans for fight, flight, freeze, or friendly (smiling, talking) reactions in the body. This limbic stimulation is what keeps us turning the page when reading a dynamic book, on the edge of our seats in a suspenseful film, and ready to flee at a startling sound. It is also what invigorates and exhausts us the first time we listen to atonal music, watch a contemporary dance concert, or walk down a busy urban street.

Artists seek the balance between what is new—stimulating limbic alertness—and what is familiar—allowing the relaxation necessary for layered association and connection. In music this might be the ease of a familiar rhythm or a melodic line amid disparate sounds, evoking a sense of participation on the journey of the unknown. Eventually, what used to be "different" becomes

During childhood, I visited the Art Institute of Chicago each year with my father. There I learned to recognize artists and epochs by their style. Once, we attended a Picasso retrospective: enormous rooms detailing one person's view. Walking by the early rose and blue periods, with paintings of Saltinbanques by the sea; strolling through the giant round women lounging on their canvases; and adjusting to cubist fragmentation, I thought: this is what it means to experience life through art.

When writing a book, there is a time in the process in which it begins to have its own life. The experience changes from leading to following. "At this point," a colleague says, "it's no longer like leading a pet; it's like tracking a wild animal." Creative projects pose questions; if we don't discover something new in the process, the work is lifeless. A builder describes a similar experience in constructing a house. "At a certain point it becomes its own creator. If you force it into something it's not, it's going to cause problems." And then he adds, "It takes a fair amount of letting go of your identity, to let it have its own energy, its own personality." A choreographer reminds me, "A dance isn't really good until it does you."

the norm, and the dynamic edge of the art form shifts to new or very old ground to heighten arousal. Art has long played a role in healing, for the artist as well as for the audience. It activates deep internal processes integrating the new with the old, the foreign with the familiar.

Creativity encourages diversity. Yet the unusual person in a family, school, or community; an odd tree or weed in a cultivated landscape; or the mysterious lone wolf or bear in the wild often are viewed as a threat to order. How do we incorporate difference in our understanding of the ordinary? It is helpful to recognize that creativity is a natural phenomenon, producing images that have meaning in our lives. The influential part of the mind that creates symbols remains largely unexplored. Our creative response to life is essential to our survival, inherent in our structure, just as the basic characteristics of life are present in each cell's capacity to reproduce, to metabolize, and to respond to changes in the environment. Human nature is reflected in our urge to create, to participate, and to respond.

Each of us needs a form to contain and transmit the deep experiences of our lives and to help us articulate and communicate our feelings and insights. We have grown accustomed to high-end art: perfect recordings, museum paintings, and multimillion-dollar films. But creativity is personal. As we participate in the process, we recognize that it is not just about observing, not just about buying and consuming, alienating ourselves from our own creativity. It is about participating in the creative universe we live in. Steven Rockefeller, biographer of the great humanist John Dewey, offers the view that art is inherently integrative. It is more than the sum of its component parts. There is magic, mystery in wholeness. In this way, creativity and spirituality are linked. Good art, as with any integrative experience, brings us directly into contact with the numinous, the unknown.

Those of us involved in environmental work particularly need avenues for expression and for celebration. Daily awareness of the environmental crisis can drain our inner resources and pollute our biological systems through despair and grief. Participating in artistic endeavors on a daily basis nurtures connection. Creative expression is a lifelong process that occurs in overlapping sequence. Part of our task as creative individuals is to decide which resources are for personal play, recovery, and healing; which are ready to be formed and communicated to others; and which are to be nurtured and contained for future expression. There is no need to burn out, to overextend our resources as individuals, just as we are depleting the earth. A creative life embraces the hard work of play, sustaining our capacity for spontaneous joy while translating our findings, our best insights and intuitions, to the larger community. Time spent in creative exploration allows recovery as well as expression, replenishes our body as well as our commitment to attend to the earth.[1]

As we feel, we care. In our busy lives, creativity helps us remain sensitized both to the complexity and to the hidden order that is beyond our rational understanding. As we open to our creative potential, we recognize we are part of the mystery.

Creative process

10–60 minutes

• Consider your creative process. When are you receptive to intuitive knowing and images: walking late at night or early mornings, listening to music? What interrupts this process for you: fatigue, too much talking, stimulants? How do you "cue" your creative imagination: do you need to sit still, move, take a walk? Where do you work best: indoors, outdoors, in a crowded café?

• Look around the space where you work. How much clutter is in the way? Empty the space of everything that is not essential. Ask your body what is useful in your creative environment and what is distracting. When you bring something new into the room, consider removing something old. Take time to remove what is keeping you in the past rather than the present.

• Create a dialogue with your process. Write about what you find.

Arranging (alone or in a group)

40–60 minutes

Outdoors, in a clearing near different kinds of terrain:

• Take time to gather a group of natural objects from plants that can be found nearby: pinecones, nuts, seeds, bark, pieces of wood, fabric, leaves, and grasses. Add stones, shells, or containers of water as related. Choose something large for the center.

• Utilize all your senses; smell, taste, listen, touch, look, move, and feel your inner responses as you make choices. 10 min.

• Begin to construct an arrangement, without speaking. Consider the objects; consider the space around the objects—allow both to be important. Take all the time you need. Include the soil in your view. 20 min.

• When you are finished, step back and notice the whole; then notice the parts. Walk around the perimeter. Where is the outside edge of your arrangement, what are the pathways leading toward it, and how do you know? Write or talk about your experience. 10 min.

• When you are ready, begin to deconstruct your arrangement, returning each of the objects to its place. Notice how your arrangement looks at various stages in the process, the ways in which memory is instilled through touch and movement. 10 min.

• Pause in open attention; experience the place as it was, before your creation.

   Indoors: Take time to arrange flowers or a group of objects. Be aware of positive and negative space: the elements and the space around them.

Stacking rocks, Herb Ferris. Photograph by David Colton.

Visits

2–4 hours

1. Attend a theater, music, or dance performance. How is your body involved; what's the relationship established to place?

2. Visit a museum or outdoor sculpture. What attracts your attention? Notice your response, including ideas, feelings, emotions, sensations, and images.

3. Visit a church, meditation center, or building that you like. What is its relationship to your body and to the place it inhabits?

Write about your observations.

Place visit: Creating place-based art

Choose a medium to work with and bring it to your place. Bring your clarinet, your sketchpad, your notebook, or your camera. Stay simple; don't get too elaborate to begin with. Establish a creative dialogue with your place. Remember, your place is not a backdrop for creative work; it is a participant, a partner, a collaborator. If you have no familiar medium, choose one to try: a poem, a story, a painting, a few minutes of movement. See what emerges; allow surprise. 20 min. Write about your experience. 10 min.

Over time: Create an artwork based on some aspect of your place. Be informed; draw on all the information you have learned. Absorb this information deeply enough that it can be integrated into an expressive medium. Be specific as you connect to soil, animals, water, air, or plants. Draw from your stories and field notes. Art is filtered through your personal experience; it is distillation that creates a poem rather than a report, a dance rather than an exercise.

Block from *River Section,* sculpture by Harriet Brickman. Cement, stone, metal.

Specifically for dancers: Site works

Ten questions for site-specific works:

1. *Attitudes:* What is your relationship to the site? Is it a background for your dancing, a collaborator in the process, or the focus of the event? (Consider place as having intrinsic intelligence with much to teach you.)

2. *Scale:* What is your relationship to time? Do you know the geological, biological, and cultural history of this place? (Research, be informed, share your knowledge.)

3. *Memory and association:* What kinds of memories and associations do you have with this place? Are you working with fantasies, dreams, emotional connections? (Acknowledge that landscape is emotional and engages empathy, projection, expectation, association.)

4. *Training:* How will you train sensitivity to place? (Consider that you are nature too.)

5. *Constructing the dance:* How do you move to and through the landscape? Walk and feel your pathways. Consider weather, angles of light, perspective. Work at specific times of day. Consider the rhythm of the topography. Clarify beginnings and endings. (Know that there will be surprises; build them into your plan.)

6. *Relationship to audience:* How will viewers interact with the site, how many people can the site accommodate, where will they sit or stand, how will they know their place and the cues to begin and end? (Help viewers feel comfortable as participants.)

7. *Performance as ritual:* Is there a ritual component? Is anything given or received in the process, as an offering or exchange? (Repetition and intention create ritual.)

8. *Finishing your work:* How will you honor the site when you are finished? (Clean up; put it back as it was. Create a conscious ending to your relationship.)

9. *Articulating the process:* What do you want to communicate in posters, press releases, programs, rehearsal process, and performance? (Find clear language for place-based work.)

10. *Evaluating the project:* What are your criteria for a successful event? (Maintain your aesthetic values; open to a larger view.)

WRITING HOME: FIRE

My core burns. For nights now I lie awake. Outdoors at dawn, which comes early in Vermont, birds call to each other with determined voices. I sit under gray skies and watch the giant poppy break through its prickly pod. Already the bearded iris are extravagant in their blossoming. Redwing blackbirds guard nests in the cattails; pale blue eggs, scrawled with brown, are hidden close to the water.

Yet I am longing for something I cannot name. Wind through branches ruffles papers and leaves, pointing the way. Yesterday I said to my stepson, "You could be offered the right school and the right job and not even know it. Instead, you would find everything wrong." Now I breathe in the meaning of these words in my own life, resisting joy with all my heart. I am in love with the man and the places I have chosen, without even knowing it is so.

Read aloud, or write and read your own story about creativity.

DAY 30

Concerning the Spiritual

[T]he element most lacking in our lives today is a sense of the sacred . . . an awareness that both body and earth are sacred.

—Matthew Fox, *Inner and Outer Ecology Conference*

ENDINGS AS BEGINNINGS

"You can't use the word spiritual in your writing until you can define it," my incisive friend remarked, shaping my relationship to this topic over the past decade. Then, teaching a course for a "Religion and the Arts" symposium, I asked the students to write their history of religion. They uniformly responded: "Religion isn't important in my life," and turned in lengthy, enthusiastic papers. I called my friend and said, "Spirituality is access to something larger than the self," and began writing this chapter. A yoga teacher offered, "Spirituality is giving more than you take."

Reflection on body and earth reminds us of our wholeness. We lose our aloneness and experience participation and community with living and other-than-living aspects of our world. This sense of oneness invites the numinous, the unknown. Spirituality can be described as that state of spaciousness and receptivity that animates the body and connects us to the mysteries of earth. Our attention is drawn to the foundations of existence and our place within the cosmos.

Spirituality is an acknowledged attribute of humanity, involving heightened awareness, spontaneity and surprise—the capacity for awe. Three qualities common to religious traditions worldwide are that they include an ethical and moral component, engage silence and inward reflection, and offer service to the larger community. Once rooted in particular landscapes, they convey knowledge of nature and culture gathered through the millennia.

However, there is a distinction between religion and spirituality. Religions encompass the rites and rituals, texts and dogmas associated with specific traditions; spirituality is a state of mind that animates religious practice. Ideally, they are so complementary that they become interwoven. In wisdom traditions throughout the ages, spirit has been identified by various names, such as Creator, Great Spirit, Higher Power, Source, Divine Mother, and God. This diversity of manifestations reflects the dialectic between form and formlessness that underlies spiritual experience. In fact, the mystical dimension of religious traditions dissolves boundaries, engaging the interconnectivity of all spiritual practice.

Diverse views concerning stewardship of body and of earth underlie our attitudes and may need to be sensitively revised in a contemporary context. A key element in that process is a reconfiguring of our relationship to religion and spirituality, drawing on ancient traditions for contemporary life. Monolithic institutions are slow to change, yet there is value to the unchanging dimension. It is critical, in fact, that the container, including methods and ideals, can withstand the test of time. Yet adaptation and responsiveness are acknowledged components of effective form. Attitudes of dominance and dominion are being replaced by concepts of unity and interconnectedness as we understand our role in the larger scheme of things.

Perhaps our deepest spiritual desire is to regard ourselves as an integral

Study for Mona and the Burning Suit—with Boat, drawing by Gordon Thorne. Oil and graphite on paper, 40 in. × 52 in.

part of nature. Many humans turn to nature to access spirituality. Through direct experience with the natural world, we engage other-than-human intelligence, powerful and mysterious qualities, and a grander scale of time. A walk in a healthy woods can slow us down, open us up, and keep us alert and responsible for our own survival. We are embedded in something larger than human concerns: protruding bedrock may be billions of years old, the old-growth hemlock trees may be hundreds of years old, and the trout lily underfoot may bloom for only a few days. Nature is immediate, dynamic, beautifully formed, and unpredictable, reflecting our own wild nature back to us. If we consider ourselves part of the natural world, we celebrate its grandeur and awesome energies; if we feel separate, we may experience fear or insecurity and the resulting desire to control or destroy landscapes and ecosystems.

The twentieth century involved a remarkable merging of Eastern and Western spiritual traditions, offering new insights concerning the unity of spirit and nature.[1] In 1893, Sri Vivekananda became the first Indian yogi to

Figure of dignitary or priest. Mexican, Tabasco, Maya 410–650 A.D. Metropolitan Museum of Art, The Michael C. Rockefeller Memorial Collection, Bequest of Nelson A. Rockefeller, 1979. (1979.206.1063).

come to America. Further manifestations in the spread of Asian religions to the industrialized world include the writings of Zen scholar D. T. Suzuki and the popular spiritual leader J. Krishnamurthi. The immigration of Buddhist monks and scholars, including His Holiness the Dalai Lama, after the 1959 Chinese invasion of Tibet, and of Thich Nhat Hanh following the Vietnam War, brought profound meditation techniques and teachings to the West. The influx of holistic medical professionals, exemplified by endocrinologist Dr. Deepak Chopra of India, brought Ayurvedic traditional practices into the mainstream in the latter part of the century, invoking deeper investigation of the spiritual dimensions of healing. Movement forms have also made their impact, including the popularity of yoga from the Hindu tradition, the graceful choreography of Sufi whirling from the Islamic tradition, and tai ji (t'ai chi) and Qi Gong from the Taoist tradition as avenues of awareness and integration of body and earth.

The American continents also reflect a dynamic spiritual heritage, newly recognized as existing prior to the overlay of Christianity and underlying it. This includes the traditions of native peoples: African American and Native American, Mexican, Latin American, South American, Eskimo, and neighboring Polynesian cultures that honor the mysteries of life. Black Elk's visionary views were documented in *Black Elk Speaks*. Principles of stewardship were embodied by the life and practice of Helen and Scott Nearing and documented in their classic text, *Living the Good Life*. Gary Snyder and a host of contemporary writers translate the Buddhist principles of mindfulness into a bioregional ethic. Thomas Berry explores the relationship between human cultures and the natural world in *The Dream of the Earth,* as well as having inspired numerous nonsectarian books that continue to focus attention on spirit and nature.

To most people, spirituality is synonymous with light and transcendence. Expressions such as "to be enlightened," "to see the light," "to receive insight" reflect the degree to which fire and light are part of our spiritual aspirations. Yet traditions worldwide also remind us that spirituality is rooted in the here and now, embracing a continuum of light and dark. We engage life as it is, not as we would like it to be, experiencing mystery with the ordinary as well as the extraordinary. In many traditions, light and dark are essential components of a whole, such as in the Taoist symbol for yin and yang. The Christian Black Madonna speaks of hidden mystery within; the Hindu Kali, of the destructive and transformative forces; and the Native American coyote and African Anansi the spider reflect our inner trickster and clown. Nature itself is multiple and complex, not simply idyllic and spectacular. The experience of the spiritual often draws us down, to the roots, the soil, and the foundations of experience. Caves, grottos, and pools are places of mystery, as are mountaintops, sky, and clouds.

Place and space shape spiritual experience. Sacred sites are identified in every culture and across cultures. Where indigenous populations are still intact, a grove of trees, a valley, a mountain, a circle of stones, a pool of water, a river, or a stream can inspire pilgrimage; a single tree, rock, cow, or bird may hold a place story of the sacred. Sacred architecture reflects earthen forms: trees, caves, pillars of stone. Some feel there are unique vortexes of energy fields that make particular sites unique; others feel that it is our intention toward a place that changes its effect. Designated space is a component of spiritual life, serving as a link to larger dimensions. Quiet space can be a rare phenomenon. Many students, adults, children, and senior citizens find it difficult to locate a space where they can be silent and take time for integration and spiritual unfolding in their workplaces, homes, preschools, retirement communities, or assisted living facilities.

A ritual is a focused container for experience. Communal or personal, the precision and repetition of ritual creates its charged effect. A simplified view of a ritual or ceremony of transformation shows three essential stages: preparation, enactment, and return, including completion and time for recovery. Thus, the end becomes the beginning, connecting to the future as well as reflecting on the past. Ritual stages also require transition from one dimension to another. This period, called a liminal state, is a betwixt-and-between time when one is neither who one was nor who one will be. It is a particularly potent and dangerous period, requiring care, teaching, and support from elders and the community. These periods of vulnerability and heightened potential occur for each of us in our journey through life.

Often we long for ceremony to ritualize transition, to welcome us home, and to recognize our new relationship to community. Contemplative practice is a nonsectarian tool, a form of pedagogy for heightening awareness and deepening experience, challenging thought patterns destructive to life on the planet. Based directly in silence and focused attention, forms of contemplative practice engage us in the moment, including inner and outer dimensions. Every religion has developed forms of contemplation to orient the mind toward spiritual concerns. Prayer, meditation, postures, chanting, prostrations, davening, spinning, dancing, and drumming are a few examples. Each offers a container for mystery, addressing experiences and fulfilling needs beyond rational understanding.

Yet contemplative techniques can be experienced separately from a religious context bringing insight into experience. Contemplative practice focuses attention on what actually is happening, rather than on ideas and habit patterns that tell us what is occurring at any moment. Cultivation of inwardness and silence, along with focused attention and nonjudgmental awareness, help us to observe patterns of mind without reacting. For example, in Vipassana meditation we attend to breath and sensation in the moment, without habitual response. This requires a safe, supportive environment so that the autonomic nervous system can relax and allow deep investigation and integration. As familiar neurological connections are bypassed, nerve endings literally grow additional dendrites, branching out to form new connections. By interrupting our habits we can create an opportunity to release old patterns, and encourage new connections.

Form in contemplative practice is often described as the banks of a river: the banks provide a container that allows us to focus on the water. Yet if all our attention remains on the banks, we may never experience the water. Thus, contemplative practice techniques are often quite simple,

When my older stepson, Josh, became thirteen, he wrote a story about fire and read it aloud at a performance. Jungian psychologist Edith Sullwold asked him for a copy. "It is a classic adolescent transformation story," she explained to me. "Burning, spinning, disorientation, emergence—a self-initiated rite of passage." When my younger stepson, Jonah, graduated from junior high, we sat through a particularly dull and uninspired ceremony. Six awards were given, all for good conduct, all to girls. He returned home, took off his shirt, collected the family canoe, and spent the rest of the day executing his first solo paddle on the Otter Creek River. When he finished at dusk, he walked up the grassy bank looking radiant and refreshed. There are so few effective community rituals; we learn to make our own. The task is to recognize the moment.

OF SPECIAL INTEREST:
TAI JI (T'AI CHI)

Doing tai ji amid a grove of white birch on an evening walk changes how I see. Following a simple form called "Opening Circle," by Chungliang Al Huang, I begin with an open stance, embracing an empty circle with my arms. Then stepping one foot back, I silently whisper "Kai Sing," opening my heart and spreading my arms. Fingertips touch the horizon line as I balance left and right, up and down, merging polarities through my core, Preparation. *I draw one leg up toward my pelvis and stir, waking up the dragon within,* Fire. *Then, shooting this enormous energy forth with both arms, I rise on my toes, stretching my whole body up toward the sky,* Air. *And then moisture rains down through me, dripping off fingertips, entering the ground,* Water. *There it is absorbed by my roots, travels up the trunk of my body, branches out into a tree,* Wood.*

A slow circle reveals the circumference of my vision, looking out from my central core. Then I pause, rooted, reaching my left arm behind my body, opening the left side, scooping up all that came before—my evolutionary origins; then the right reaches back, scooping my own life history and bringing it to the front, where I can see clearly. Both hands meet at my heart, where the energy is distilled, crystallized, Metal. *Then all drops away, released (swoosh) to the earth, followed by a moment of emptiness, a gathering of that which is essential.*

their philosophic underpinnings communicated through simple stories as well as complex texts. We don't want to be distracted by the outer form, the detailed trappings of the technique. In some ways, the simpler the form, the more potent the experience. In Vipassana meditation, for example, the form is concise. The body remains still (usually in seated posture), and we focus on breath and sensation. In the movement component of yoga (hatha yoga), the body engages specific postures as we focus on breath and sensation. In the discipline of Authentic Movement, the body follows deep inner impulses for movement, and an external witness holds the role of the conscious, observing mind. In the flow of tai ji, the outer experience is repeated, but the inner experience is new each time.

In each technique the role of the mind is to remain nonjudgmental but discerning, noticing without reacting, following the sensations of the moment without spinning off into past or future. Experiencing the edge between new and old without strain, eyes often engage soft or inner focus to heighten awareness of internal states. The role of the practice itself is not to retreat from society; it is to prepare us for engagement, "bringing our full selves to the table" in any situation. Repetition of form can deepen understanding. As we "practice" contemplation through repetition, the form becomes less and less important. The complex fingering of a violinist or the polyrhythmic facility of a drummer are skills that require practice for freedom of expression. The same is true of form in contemplative traditions; mindful repetition offers a gateway to much larger experience.

Heightening awareness of sensation allows us to track experience—and our response to experience—directly through the body, moment by moment. If we see a person wearing a green shirt walk into the room and we begin to feel heat in our body, we can notice through our sensations that we are reacting to that person with anger or alarm. We may not even consciously know what we are feeling. Perhaps we were once treated badly by someone in a green shirt. But we don't need to know *why* we have that reaction; once we are aware of our subconscious response, we can breathe deeply, calming our autonomic nervous system, and take time to notice if, in fact, there is cause for alarm. Then we can make choices about what to do. It is this moment of pause—or breath—between perception and response that creates choice in our lives.

Another example of the role of the senses in contemplation might occur when we sit down at a big meal and dive into our food, eating automatically. Instead, we can take time to allow our senses to register the smell, the colors, and the visual aesthetics, including the environment in which we are eating. In this way, each bite registers in the body, nourishing us at many levels. We may be reminded of the interconnections between body and earth, the history of the food and those who helped produce it. The process of eating becomes less about consuming and more about participation in the vast and intricate universe we live in. As we engage contemplative mind, we confront the fundamental delusion of separateness. Any moment can be filled with awareness if we allow our body to fully register—take in, associate with, and respond to—the experience at hand.

Contemplative Practice in Education

Contemplative practice balances inward reflection with outward expression. By doing contemplative forms, such as meditation, yoga, and authentic movement, we are participating in knowledge, immersing ourselves in a process of inquiry rather than acquiring knowledge, which is a consumer analogy. Contemplative forms connect us to history, to educational traditions that have been refined for thousands of years. Distinct from particular religious orientation, contemplative practice includes joy in the life of the embodied mind, offering an opportunity for discovering who we are and what we think and believe, especially in relationship to the earth.

Today much of our contemporary Western thought process and educational pedagogy is oriented toward productivity, personal enrichment, and self-enhancement. Integration through unifying experiences is often lacking, reinforcing fundamental delusions of separateness and dominance. Contemplative traditions, in contrast, engage place-based values and heighten awareness of self and community. (For example, in the discipline of Authentic Movement, when an individual mover rejoins the circle of witnesses, relationship to the collective container is heightened.) Contemplative traditions help us to focus attention in the moment, to acknowledge and release unwanted patterns from our past history that may block our attention, and to invite experiences of wholeness and participation as we envision a future—inspiring hope.

Contemplative traditions extend attention to a larger community, one that includes the interconnected web of all living and other-than-living components of the earth—the humans, animals, plants, soil, rocks, air, and water around us. This simple, fundamental shift in attitude changes what will and can occur. In this way, contemplative practice provides a context to help us reorient how we think and live in terms of ecological values. It also extends attention to broader dimensions within ourselves, acknowledging the dynamic inner landscape of thoughts, emotions, sensations, and intuitions. By including both the internal and external aspects of our daily experience, contemplative practice challenges our habits and awakens fresh response. Thus, forms of contemplation create a mental environment conducive to larger moral and intellectual concerns and a physical environment respectful of body and earth.

As we come to the end of our map through this text, we can consider the pilgrimage we have taken together: journeying to the origins of life and paying homage to ancient ancestors and traditions that support our contemporary lives. As we return to our communities and to the places we inhabit, we can acknowledge the spiritual dimension—the unknown—that links body and earth, informing all that we do.

❖ ❖ ❖

A final circle of the arms brings me to Embrace Tiger, *letting the animal tail slip through my hands without grasping, feeling my own wildness, as I* Return to Mountain, *acknowledging this place where I stand. The surprise is that I experience the landscape differently when I finish. Today, I notice two downy woodpeckers etched in the top of a white pine, silhouetted against the evening sky. I hear the whisper of wind through the birch leaves, like a voice, and admire tiny mushrooms an inch from my foot. They were there all along, but I wasn't prepared. Moving in nature allows me to enter the landscape, become part of all there is to see.*[2]

Yoga: Downward Facing Dog Pose, Adho Mukha Svanasana.

Ritual circle
15 minutes

Standing outdoors in a field, eyes open: Begin walking in a large circle. Use inclusive attention as you walk, remaining aware through your senses of both body and place. Keep circling, until you feel the integrity of the circle established in this place; it may take five or six cycles. Reverse directions.

• Then pause and face into the center of your circle. Acknowledge all that is inside. Imagine something from your past that you would like to leave behind you, inside the circle.

• Turn and face outside the circle. Acknowledge all that is outside. Imagine something from your future that you would like to move toward. Visualize how your body would look and feel. Now walk away from the circle, allowing your movement to be informed by your image; turn and stand as you would be standing.

• Notice what you have made, what you have left behind, what you intend for the future.

• Pause and enjoy the moment.

The discipline of Authentic Movement: Triads
60 minutes

Working with the discipline of Authentic Movement, indoors (in a comfortable movement studio), grouped in threes: one person as mover, one as witness, one to witness the whole (including mover, witness, and the space—holding the field):

• The group begins seated in a large circle around an empty space for movement: the designated movers (of each triad) enter the movement space, close their eyes and begin following inner impulses for authentic movement—a process of body listening. The witnesses (one from each triad) continue sitting in the circle, a nonjudgmental presence inclusively attending to what is being seen and what is being felt. The holders of the field (one from each triad) sit outside the circle, at the outer rim of the space, and witness the whole. After 10 minutes, a designated witness calls the movement to a close. The movers return to the witnesses, joined by the "holder of the field" and sit together in open attention.

• Then change roles and repeat the process. (The holder of the field witnesses, the witness moves, the mover becomes the holder of the field.) 10 min.

• Change roles once again, until each person has experienced each phase of the experience. Take time in the triads at the end to draw, write, or speak about the experience.

This is an introductory description of the triad form; it varies according to depth of experience. The main thing to remember is that the mover remains the expert on her/his own experience; speaking helps reflect back the multiple perceptions, but it is not the focus of the work. The intention is to be seen and to see, without trying to "do" anything. In this context the body begins to unfold its movement language, including access to spiritual dimensions.

Place visit: Attention to the spiritual

In tai ji receptive stance (legs slightly bent, arms encircling the space in front of your chest, vertical spine, soft, responsive belly):

• Begin moving with the intention of honoring the earth; remember this movement or sequence of movements.

• Begin again, repeating the movements for earth and adding a new movement honoring air. (Remember, you can turn, change levels, move around your base.)

• Begin again, honoring earth and air and adding a new movement for water.

• Begin again, honoring earth, air, and water and add a new movement for plants.

• Begin again, honoring earth, air, water, and plants and add a new movement for animals.

• Begin again, honoring earth, air, water, plants, and animals and add a final movement, uniting the whole.

• Repeat the whole sequence several times, keeping movement and intention clear. You have created your own movement ritual, a form that can be repeated at other times and places to honor the earth. 20 min. Write about your experience. 10 min.

the lush green of Vermont. This pastoral landscape disguises a land twice emptied of trees and of people in the past two hundred years.

You always know you are in Vermont because people are outdoors doing things. Larry's Campground is full, a father and son finish mowing a field, a woman loads a cello into her car. As we climb up the rounded 350-million-year-old Green Mountains, a woman walks down carrying a rock on her head. At the top we gaze over the beaver pond below. Silos in the valley tell of fertile soil, twice a sea, twice a lake.

I am prepared for transition but not for arrival. In less than two weeks I will teach students from all over the world—every state, mostly cities—many of whom have never stood beneath a white pine before. They will stare at corners of rooms, qualities of light, seeking the familiar. In this season of migration, we each need landmarks detailing the journey as we begin to inhabit this landscape as home.

Read aloud, or write and speak your own story about spirituality.

DAY 31

Embracing Place

There is nothing of me that is not of earth, no split instant of separateness, no particle that discriminates me from the surroundings.

—Richard Nelson, *The Island Within*

GROWING

When a friend was getting remarried, her son passed me on the steps in a flush of excitement. "I just don't know if my heart is big enough to hold another person!" he exclaimed. When I was getting married, my younger stepson asked his father, "Does that mean I will have two grandmothers?" "Yes," my husband replied; "maybe three, maybe four." "Good." He grinned back. "Then both cheeks will be red from getting kissed so much!" Once I stood on a lake shore, so full of joy that I stretched my arms wide, pressed my breastbone against the warm pulse of air, knowing earth as body.

Humans inhabit the insides of things. We can think of earth as surface, a topography amplifying the curves and valleys, sensuosity and vulnerability, bold contours, erupting cavities, and unstable textures of our own skin. Going below the surface of both body and earth, we find the heat that animates and keeps things moving. As we imagine earth as body, we can no longer remain separate. We recognize that we are utterly dependent on and interdependent with earth systems. That is not cause for despair but for celebration.

Place is the medium through which we experience the larger dimensions of the earth. Even a small, natural place is as comprehensive in scale as a body tissue or organ. The shaded area under the hydrangea bush, where we used to play hide-and-seek at dusk, is an intricate ecosystem with all the inherent components of earth, air, water, plants, animals, and fire—internal and external energy sources. This cool, sun-speckled terrain has geological, biological, cultural, and personal history embedded in its soil. We can contemplate what it looked like fifty years before our arrival, one hundred years ago, one thousand, one million, or one billion years ago. Or we can imagine how it might be in the life span of our children's children. As we explore and attend, it gets larger, not smaller.

A variety of plants, animals, and living organisms inhabit healthy places. Moving through such environments, we experience an intricate web, a conversation of purpose, a cacophony of scale—sound, tactile and visual components that create a larger whole; it's more than the sum of its parts. Diversity requires more of us, drenching us in the richly textured moment. A neat lawn, for example, is less complex than a wild field; the mowed edge of the freeway, less than the waving grasses and wildflowers beyond. Creating a monoculture, often through herbicides, pesticides, and vast expenditures of water, lawn-keepers (those in love with perfect lawns) perpetrate the same green, season after season, out of step with larger cycles, including indigenous flora and fauna.

Place has dimension. This becomes evident as we invite our senses to open to increasingly subtle layers. What appears solid becomes a matrix of ever changing relationships. For example, on a first visit to the Grand Canyon, it is tempting to take a photograph and turn away, because the magnitude of

the one-mile-deep crevice requires such a stretch of perception. Hiking down a zigzagging path from the north rim, descending in time, we see other layers emerge. Exploring the side canyons, swimming in their clear streams, and sleeping by the Colorado River at its core, we notice unexpected dimensions unfold; and eventually the canyon moves inside us, affecting the rhythms of breath, pulse, and the weaving pattern of our footsteps. Perception is a process, not an event.

Understanding place requires participation. Although the Grand Canyon is spectacular, we can sit by any healthy stream and invite depth perception. Take off both shoes and wade in, feeling the textures of rock; lift up a stone and watch for insects; look closely at plants and animals; explore the bank and have lunch, then relax, digesting all we have taken in, allowing associations and connections. We can spend a lifetime getting to know a stream. Home places, too, require active attention. We select places to inhabit that reflect our inner terrain. Our choices involve gut response as much as conscious planning. We know what we need around and within us, including what color to paint the walls, which plant to place by the doorway, and how

Today I received a letter from a friend. Years ago we slept in fields under open skies, passed through the cosmic convergence, stood amid old-growth redwoods, and ate blueberry pancakes in a garden filled with flowers. We also talked of honesty, the nature of emotions, and a life devoted to art. I am fortunate to have had several friends who were willing to engage such depths. Each time I perform or write, I draw on all we have shared. Now, as I near completion of this book, I walk down a road in Maine to where I began and stand gazing at a pale pink sunset at the crest of Dogherty Ridge. It is useful to recognize endings as part of the continuity of life. My eighty-two-year-old-mother is driving on logging roads with a friend, looking for moose. It all feels familiar: opening, filling, offering, emptying, and beginning again.

Housepost figure. Primitive, Oceanic, Polynesia, New Zealand. H 43 in. The Metropolitan Museum of Art, The Michael C. Rockefeller Memorial Collection, Bequest of Nelson A. Rockefeller, 1979. (1979.206.1508). All rights reserved, The Metropolitan Museum of Art.

to configure the path to the garden—without charts and books encouraging conformity. These constructed patterns form their own language, like the earth body itself, that affects our days.

Place involves relationship. As we explore, commit, and remain present over time, we bump into our own limitations. In a consumer economy, it's all too easy to use the one-night-stand approach to place: visit and leave. It is often easier and sometimes necessary, as our lives move us from place to place, to remain detached and pretend distance, to disregard and deny connections. Instead, as we commit ourselves to place, it unfolds its wonders. We begin to notice the vole moving right under our feet, the annual passage of the yellow throated warbler, the auspicious appearance of a young oak seedling, planted by squirrels, that will grow, as we grow, larger and more solidly embedded in the land through the years. Place changes, just as we change.

Place evokes emotions. Where did we first feel safe and at home, where were we fearful? Emotions become wired into our neurological network in the early years of our lives, linking experience with sensation (feelings) and language. Perception of place merges the exterior landscape with powerful interior responses. Some of us are fortunate to find at least one place on the earth where the landscape feels as familiar as the palms of our hands, informing us about underlying connections. When places "feel right," they are extensions of ourselves. These places offer comfort, challenge mediocrity, and evoke our "best selves." Sometimes, absorbed in the romance of place, we just want to visit these sites, rather than make them our homes. Place as commonplace might be too much. Yet as we recognize the many ways that place affects us, we can make choices about how and when to engage with it and how best to care. Sometimes, when we are most at home in place, opening our hearts, and speaking freely and honestly, we are also most vulnerable. Deeper layers of experience move to the surface with surprising ferocity.

Place evokes grief. For many of us, memory is associated with specific terrain: the quality of light when we fell in love, the room where a child was born, or the tree climbed in moments of retreat. All too frequently, loved ones are gone, places paved over—in reality or in our hearts. Grief is the natural response to loss. Our reluctance to grieve can interfere with our capacity to feel. Even beyond the personal currents of our lives, earth systems are dying. Forests, whales, ponds, and the air in our lungs are in trouble. No wonder we suffer as individuals and as a culture from grief and depression. As grief is repressed, its companion symptom—obsessive consumption—fills our days, even against our best efforts. Yet grief opens us to the truth of what might perish or is perishing.

Curiously, humans seem to value more deeply that which they may lose. The awareness that grief opens has the potential to clarify our vision and move us to respond. "Denial of our own mortality is intricately interwoven with our denial of the earth," writes grief counselor Pat Noll.[1]

We relate to life-threatening illness as though health means that life never ends, denying relationship to larger cycles. If we refuse to acknowledge our own potential to die, we will also blind ourselves to the vulnerability of this earth that sustains us. Individuals consciously engaging the end of their lives can teach us about changing attitudes toward earth and body. Confronted with the reality that treatment will no longer cure, they may work to reconceive their lives, reaching out to connect and cherishing what remains. Grieving what is lost, they—and we—can often embrace life not for its duration, but for its essence, here, now, today.

Health of the body is reciprocal with health of the earth. We need to be vigilant about what we feed the landscape of home, as well as what we feed our bodies. Strip malls and highways covering the earth are cancers to the earth body, threatening essential life processes. In fact, the basic definition of cancer is "uncontrolled growth." As we care for and embody ourselves, we are contributing to the health of the earth. As we shrink, shrivel away, and participate in the fragmentation so prevalent in our culture, we reduce the health of our place. Earth is body, and humans are lively, functioning, often threatening components of the whole. Our attitudes, images, and visions come to bear on the life systems in which we are embedded. As we participate in the dance of place, we partner the magnificent efficiency, the spectacularly adjusted systems, and the exquisite choreography of earth.

Our tendency to idealize or romanticize nature is as dangerous as uninformed fear. We begin by listening to the earth, using the resources of the body. As we attend to underlying patterns, we recognize that body is part of earth—inseparable. Each individual is not alone but a participant in local and global bioregions, just as every cell is part of us. Through place, we can focus rapt attention on one particular site, opening perceptual gateways and softening our habits of projection and association. We engage our multiple intelligences through writing and story telling, drawing and visualizing, moving and making, listening and vocalizing, cooperating and connecting, and thinking and feeling. This connection to place involves being who we are (rather than who we might construct ourselves to be) in the grander scheme of things.

Humans live somewhere between the cool perimeters of sky and the fiery core of earth. The more we reach into ourselves, the closer we are to the ground, to soul. But we also must reach out, stretching ourselves toward the sun, moon, and stars, merging with the inspirational qualities of air, of spirit. Infusing our consciousness with close attention to land and place on our paths toward wholeness, we find our full dimensionality. Once we envision ourselves inside a living system, not outside, we recognize our participation in the events of the world. It is no longer a question of paying attention to the earth; it is a process of living our relationships with consciousness.

✦　✦　✦

Many times I have dreamed of following a woman with dark, flowing hair across a narrow isthmus of rock. She is familiar; I know her name. We are high in the mountains, and the precipice is formidable. The land linking one point to another forms a bridge, just wide enough for footsteps. Once across, we sit and gaze over the valleys and sharp crevices below. But always, I return, as she remains. Now, I practice staying, finding my legs, and settling my spine for the long haul. In my imagination, I look back on a life lived and left with no remorse, only appreciation. The new view crystallizes in my gaze moment by moment, dizzying in dimension. At some point, my companion disappears, and I sit, part of the rocky mountain outcrop, part of earth.

In my life, I have gone from painting to dancing to writing as my creative forms. Each provides an edge, an expansion to understanding. I painted to explore emotions, danced to embody the feelings, wrote to articulate my findings. Then there was a point in writing when I returned to dance to edit my words. As I moved, I found images so poignant I wanted once again to capture them in paint. It is all a cycle. Throughout, there are times when I simply need to quit making and go outside. Attention to body and earth reminds us that we are part of a creative universe, complete just as we are.

TO DO

Personal project

2 weeks

Create your own chapter to add to this book:

• Consider something about your home place and your body that you would like to explore in more depth. If you have trouble choosing, try a place scan and a body scan; or take a walk, noticing what draws your attention.

• Research your topic, deepening understanding. Focus on relationships between subjects rather than isolated facts. Use several texts, as well as personal investigation.

• Write short anecdotes or performance texts about your topic. Write to discover as well as to articulate what you know.

• Create your own movement exercise and add it to your investigation.

• Find an art image that connects to your topic, evoking intuitive connections. Look for a multicultural image to broaden your worldview.

• Share your findings with someone else or a group: present information, read your anecdotes, teach an exercise, and show your art image. Endings become beginnings through return to community, planting seeds for the future.

View from Tom Peevy's Truck. Photograph by Bill Arnold.

Place visit: Create an ending ritual

Create an ending ritual. Consider when you first found your place. What has happened during your place visits, and what are you experiencing at this moment? Then follow your impulses for honoring the end of this cycle of connection. Even if you plan to continue your visits, take time to mark this transition. 20 min. Write about your experience. 10 min.

WRITING HOME: HOME

Vermont is where my family was raised. Newlyweds became adults, boys became men, the newborn pup became hunter and companion. Shiny floors are now scuffed passageways marked by boots, claws, tipped-back chairs. Windows are propped open, doors stand ajar, handles no longer fit in their locksets.

In these eight years we have come to know the landscape of home in every season, hearts stretched further than we ever knew would be needed. Yet this house, still known as the Fletcher Farm, has a well so deep we can't even know its source. And each day we fill our cups and drink.

Read aloud, or write and read your own story about place.

NOTES

Introduction

1. In this text, *earth* is used as an inclusive term, referring to all aspects of the dynamic planet. Humans are part of earth. I have chosen a small *e* to reflect this intent. In Day 5, "Underlying Patterns: Earth," a capital *E* is used, designating the name planet Earth. In quotes and references by other authors, capitalization varies.

2. The phrase "our bodies" is used to imply our collective form as *Homo sapiens*. Humans share the same structure, although we inhabit this structure in very different ways.

Day 3. Underlying Patterns: Body

1. Body-Mind Centering is a registered service mark (R), and BMC (SM) is a service mark of Bonnie Bainbridge Cohen. For more information see Cohen's *Sensing, Feeling and Action: The Experiential Anatomy of Body-Mind Centering* (Northampton, Mass.: Contact Editions, 1993), 98–118, 122–55, or contact The School for Body-Mind Centering, 189 Pondview Drive, Amherst, MA 01002-3230. (413) 256-8615. BMCSchool@aol.com *or* www.bodymindcentering.com

2. Adrienne L. Zihlman, *The Human Evolution Coloring Book* (New York: Harper Perennial, 1982), 11.

3. This section was written in conjunction with Gale Turner, practitioner and teacher of Body-Mind Centering, performer and assistant director with Meredith Monk, and co-director of In-Motion Movement Therapy.

4. See Donna Farhi, *Yoga Mind, Body and Spirit: A Return to Wholeness* (New York: Henry Holt, 2000), 34.

5. Contact Caryn McHose at Resources in Movement, RR 1 Box 195 Ashland, NH 03217; (603)968-9585; www.resourcesinmovement.com.

Day 4. Underlying Patterns: The Upright Stance

1. See Andrea Olsen, *Bodystories: A Guide to Experiential Anatomy* (Barrytown, N.Y.: Station Hill Press, 1991/98), 28.

Day 5. Underlying Patterns: Earth

1. National Audubon Society, *The Sun and the Moon* (New York: Knopf, 1995).
Also note capitalization of *Earth*, referring to the name of the planet.

2. *Dawn of Humans* (Washington, D.C.: National Geographic Society, Cartographic Division, February 1997).

Day 6. Underlying Patterns: A Bioregional Approach

1. Primary sources for this chapter include Charles Johnson, *The Nature of Vermont* (Hanover, N.H.: University Press of New England, 1980/1998); Bradford Van Divers, *Roadside Geology of Vermont and New Hampshire* (Missoula, Mont.: Mountain Press, 1987/1995); Christopher McGrory Klyza and Stephen C. Trombulak, *The Story of Vermont: A Natural and Cultural History* (Hanover, N.H., 1999); Peter Alden and Brian Cassie, *National Audubon Society Field Guide to New England* (New York: Knopf, 1998); and John Elder, *Reading the Mountains of Home* (Cambridge, Mass.: Harvard University Press, 1998), integrating geological, biological, and cultural views.

2. Dates and descriptions for geological events vary, interpreted in various contexts. In this text, dates offer general guidelines to root our imagination in time.

3. See Klyza and Trombulak, *Story of Vermont*, 13.

4. Charles Johnson writes, "In fact, as the supercontinent Pangaea is reconstructed, the Appalachian system continues even farther, lining up perfectly with the Caledonian range of Scotland and Scandinavia, and exhibiting the same folding patterns and rock types as its European relatives." *The Nature of Vermont*, 27.

5. Native Americans used these waterways extensively in their migrations to summer and winter hunting and fishing grounds, paddling swiftly in their birch bark canoes. With the onslaught of European settlers, they would travel south on the Otter Creek to attack the colonies in Massachusetts (whose leaders had decimated their population by distributing blankets infected with smallpox), then return north, escaping swiftly with the current. Anyone who followed would have the water against them on return.

6. Barry Lopez, writing about the deserts of Southern California, states, "Weapons and tools discovered at China Lake may be thirty thousand years old; and worked stone from a quarry in the Calico Mountains is, some argue, evidence that human beings were here more than two hundred thousand years ago." "The Stone Horse," in *Crossing Open Ground* (New York: Scribner's, 1978/1988), 1.

7. See Johnson, *Nature of Vermont*.

8. Ibid.

Day 7. Mind: Brain and Nervous System

1. *Webster's 9th New Collegiate Dictionary*, 1991.

2. David Shier, Jackie Butler, and Ricki Lewis, *Hole's Essentials of Human Anatomy and Physiology* (Boston: McGraw-Hill, 1998), and Gerard Tortora and Sandra Reynolds Grabowski, *Introduction to the Human Body*, 5th ed. (New York, John Wiley and Sons, 2001).

3. See Joel Davis, *Mapping the Mind: The Secrets of the Human Brain and How It Works* (Secaucus, N.J.: Carol Publishing Group, 1997).

4. Shier, Butler, and Lewis, *Hole's Essentials of Human Anatomy and Physiology*, 237, states: "The left hemisphere is dominant in 90% of right-handed adults and in 64% of left-handed ones. The right hemisphere is dominant in 10% of right-handed adults and in 20% of left-handed ones. The hemispheres are equally dominant in the remaining 16% of left-handed persons."

5. The triune brain model was begun in 1952 and introduced in 1970 by brain researcher Dr. Paul MacLean. Contact Maresh Brainworks, 3505 23rd Street, Boulder, CO 80304 (303) 545-2259. Also see Paul D. MacLean, *The Triune Brain in Evolution* (New York: Plenum, 1990); and LeDoux, *The Emotional Brain*, 92–103.

6. Wynn Kapit and Lawrence Elson, *The Anatomy Coloring Book* (San Francisco: HarperCollins, 1993), 8.

7. For brain development, see Davis, *Mapping the Mind*.

8. Dharma Singh Khalsa, with Cameron Stauth, *Brain Longevity: Regenerate Your Concentration, Energy and Learning Ability for a Lifetime of Peak Mental Performance* (New York: Warner Books, 1997).

9. It seems that individual memories are stored in the cortex by the very neurons that first processed the information; emotional memories are processed through the amygdala. For various views on the physiology of emotions and the limbic region, see Daniel Goleman, *Emotional Intelligence: Why It Can Matter More Than IQ* (New York: Bantam, 1995); Joseph LeDoux, *The Emotional Brain: The Mysterious Underpinnings of Emotional Life* (New York: Simon and Schuster, 1996).

Day 8. Visceral Body

1. Natalie Angier, "Illuminating How Bodies Are Built for Sociability," *New York Times* April 30, 1996, Science sec. The polyvagal theory was developed by Dr. J. Porges.

2. Kapit and Elson, *Anatomy Coloring Book*. "Communicating rami are myelinated, so they are white" (151).

3. Ibid., 153. As the brain stem and hypothalamus receive sensory impulses from the visceral organs (by means of vagus nerve fibers), they employ autonomic nerve pathways to stimulate motor responses in various muscles and glands.

4. Angier, "Illuminating."

5. Ibid.

6. See Sandra Blakeslee, "*Complex and Hidden Brain in the Gut Makes Cramps, Butterflies, and Valium*," *New York Times* January 23, 1996, Science sec. Also refer to Michael D. Gershon, *The Second Brain* (New York: HarperCollins, 1998).

7. See Rupert Sheldrake, *Dogs That Know When Their Owners Are Coming Home* (New York: Three Rivers Press, 1999); and Carl Jung, *Man and His Symbols* (Garden City: Doubleday, 1964).

Day 9. Perception

1. Described by Bonnie Bainbridge Cohen in her book, *Sensing, Feeling and Action*, 63. In her view, the physical process of perception is at least a four-step activity. "We begin by *focusing*, or attuning our sensory organs to receive input. The act of *sensing*

involves receiving sensory stimuli through neurological receptors that exist throughout the body. *Perceiving* is the transformation of sensory stimuli into recognizable patterns of information and the whole-body interpretation of this information. *Action*, is the response—which may be to not take action."

2. This information is from William Keaton's research on avian navigation at Cornell University in the 1960s. See Stephen J. Bodio, *Aloft: A Meditation on Pigeons and Pigeon-Flying* (New York: Lyons and Burford, 1990), 36.

Day 10. Touch

1. See Davis, *Mapping the Mind*, 96.

2. See David Abram, *The Spell of the Sensuous: Perception and Language in a More-Than-Human World.* (New York: Random House, 1996).

3. Juhan, *Job's Body*, 24. "It [skin] contains six hundred and forty thousand sensory receptors linked to the spinal cord by over half a million nerve fibers, with seven to one hundred and thirty four tactile points per square centimeter of skin."

4. Frank R. Wilson, *The Hand: How Its Use Shapes the Brain, Language and Human Culture* (New York: Pantheon, 1998).

5. Juhan, *Job's Body*, 43, also states: "In two separate studies in 1915, 90–99% of abandoned infants in orphanages in ten cities died in their first year due to lack of touch."

Day 11. Movement

1. See Janet Adler's essays, chapters 12, 13, and 14 in *Authentic Movement, Essays by Mary Starks Whitehouse, Janet Adler and Joan Chodorow*, ed. Patrizia Pallaro (London and Philadelphia: Jessica Kingsley, 1999). Whitehouse pioneered in focusing attention on deep, unconscious impulses in the body. Adler brought the awareness of such work in the body into conscious relationship to a witness, grounded in her studies with Whitehouse, who was her first witness. Also see Adler's *Arching Backward: The Mystical Initiation of a Contemporary Woman* (Rochester: Inner Traditions, 1995).

2. Arlene Croce, "Balanchine," the *New Yorker* (1978).

3. Andrea Olsen, "Notes on a Sense of Place in Dance," *Orion* (New York: Orion Society), Autumn 1997.

Day 12. Sound and Hearing

1. Jeff Goldberg, *Seeing, Hearing and Smelling the World*, Howard Hughes Medical Institute (Chevy Chase Howard Hughes Medical Institute, 1995), 35. Due to "loud noises, disease, heredity, or aging—people tend to lose 40 percent of their hair cells by the age of 65. And once destroyed, these cells do not regenerate."

2. Zihlman, *The Human Evolution Coloring Book*.

3. Davis, *Mapping the Mind*, 107.

4. For information on loons, see Judith McIntyre, *The Common Loon: Spirit of Northern Lakes* (Missoula, Mont.: University of Minnesota Press, 1988), and Tom Klein, *Loon Magic* (Minocqua, Wisc.: Northword Press, 1985, 1989).

5. Ibid.

6. Shier, Butler, Lewis, *Hole's Anatomy*, 272.
7. Davis, *Mapping the Mind*, 183, 186.
8. Zihlman, *Human Evolution Coloring Book*.

Day 13. Vision

1. Keeton and Gould, *Biological Science*, 4th ed. (New York: Norton, 1986).
2. Davis, *Mapping the Mind*, 111.
3. Kapit and Elson, *The Anatomy Coloring Book*, 155.
4. See Davis, *Mapping the Mind*, 109-125.
5. Ibid.
6. Sandra Blakeslee, "Seeing and Imagining: Clues to the Workings of the Mind's Eye," *New York Times*, Science sec., Aug. 31, 1993.
7. See Oliver Sacks, *A Leg to Stand On* (New York: Harper-Collins, 1993).
8. Blakeslee.
9. Theodore Roszak, Mary E. Gomes, and Allen Kanner, *Ecopsychology: Restoring the Earth, Healing the Mind* (San Francisco: Sierra Club Books, 1995). Also see Laura Sewall, *Sight and Sensibility: The Ecopsychology of Perception* (New York: Tarcher/Putnam, 1999).

Day 14. Bones

1. For details on bone structure, see Juhan, *Job's Body* or Kapit and Elson, *The Anatomy Coloring Book*, 17.
2. For detailed information on each area of the skeleton, see Olsen, *Bodystories*.
3. For detailed information on John M. Wilson's Arthrometric Model, see John M. Wilson, "A Natural Philosophy of Movement Styles for Theatre Performers" (Ph.D. diss., University of Wisconsin–Madison, 1973), 120-22 and Olsen, *Bodystories*, 113-15.

Day 15. Soil

1. Hans Jenny, "A Life with the Soil," *Orion* (spring 1992). Also see William Bryant Logan, *Dirt: The Ecstatic Skin of the Earth* (New York: Riverhead/Putnam, 1995).
2. Research assistance, Eve-Lyn Hinckley and Greg Carolan.

Day 16. Breath and Voice

1. Kapit and Elson, *Anatomy Coloring Book*, 91.
2. Olsen, *Bodystories*, "Sound and Movement."
3. Cohen, *Sensing, Feeling, and Action*.

Day 17. Air

1. See Diane Ackerman, "Skin" in *A Natural History of the Senses* (New York: Random House, 1991).
2. Vincent Schaefer and John Day, *Atmosphere* (Boston: Houston Mifflin/Peterson Field Guides, 1981).
3. Ibid., 155, 167.
4. Schaefer and Day, *Atmosphere*, 4, 5.

5. Alden and Cassie, *National Audubon Society Field Guide to New England*, 60-61.
6. Ibid., 56.
7. Ibid., 62, 63.
8. Ibid., 65.

Day 18. Muscle

1. Juhan. *Job's Body*, 266.
2. See Kapit and Elson, *The Anatomy Coloring Book*, 9, 36.
3. See David Gorman, *The Body Moveable* (Guelph, Ontario: Ampersand Press, 1981) for excellent illustrations and descriptions of these muscles, often omitted from physiology texts. In standing postural alignment, the origin of the longus capitis muscle is on the first, second and third cervical vertebrae (C-1, 2, 3), and the insertion is on the front of the occipital foramen of the skull. The origin of longus colli is on thoracic vertebrae, T-3, 4; and the insertion is on C-1, 2, 3.

Day 19. Animals

1. Research assistance Jean Andersson and Alexa Gilbert. See Adrienne L. Zihlman, *Human Evolution Coloring Book*.
2. William H. Amos, *Wildlife of the Rivers* (New York: Harry N. Abrams, 1981). Some carnivores are opportunists, hunting whatever comes their way; others are more selective. All carnivores, however, have one thing in common: they are meat eaters and hunters, and they represent the very peak of the food pyramid.
3. David Peterson, *Heartsblood, Hunting, Spirituality and Wildness in America* (Washington D.C.: Island Press, 2000), 41.
4. Rupert Sheldrake, *Dogs That Know When Their Owners Are Coming Home* (New York: Three Rivers Press, 1999).

Day 20. Endocrine System

1. Doc Childre and Howard Martin, *The Heart Math Solution* (San Francisco: Harper, 1999). "Considered an independent entity, the heart's brain is composed of an elaborate network of neurons, support cells and neurotransmitters which enables it to process information, learn, remember, and then transmit information from one cell to another, including emotional information." (Bill Strubble, "A Mind of Its Own," *Massage & Bodywork* [Apr/May 2001], 46).
2. See Linda Hartley, *Wisdom of the Body Moving: An Introduction to Body-Mind Centering* (San Francisco: North Atlantic Books, 1989).
3. See Gerard Tortora and Sandra Reynolds, *Introduction to the Human Body: The Essentials of Anatomy and Physiology* (New York: Wiley, 2001).
4. Ibid.
5. Dianne M. Connelly, *Traditional Acupuncture: The Law of the Five Elements* (Columbia, Md.: Centre for Traditional Acupuncture, 1979), 6-7.
6. Kristen Linklater, *Freeing The Natural Voice* (New York: Drama Book Specialists, 1976).

Day 21. Insects

1. Research assistance, Greg Carolan. See the writings of Harvard entomologist E. O. Wilson, including *The Diversity of Life*. Because of the enormity of numbers involved, insect figures vary according to source. Describing the Goliath Beetle of Africa, for example, the American Museum of Natural History's display label states that beetles represent nearly one-sixth of all organisms on earth. The museum's insect collection comprises 17 million specimens representing 300,000 species.

2. Earthworms and leeches are in the phylum Annelida; slugs, snails and clams are in the phylum Mollusca.

3. North American figures from Donald Borror and Richard White, *Insects* (New York: Peterson Field Guides), vii, 45.

4. Robert Metcalf, *Destructive and Useful Insects* (New York: McGraw-Hill, 1993).

5. Wilson, *Diversity of Life*.

6. Borror and White, *Insects*, 38–41.

7. Borror and White, *Insects*, 33.

8. Ibid., 42.

9. Alden and Cassie, *National Audubon Society Field Guide to New England*, 198–230.

10. Jenepher Brettell, "The Insecticide Issue," *Vermont Institute of Natural Science's Newsletter*, March 1, 1991.

Day 22. Digestion and Nutrition

1. See Kapit and Elson, *The Anatomy Coloring Book*, 108–10.

2. See Cohen, *Sensing, Feeling and Action*.

Day 23. Plants

1. Anthony Huxley, *Plant and Planet* (New York, Viking, 1990). Also see Huxley, *Green Inheritance: The World Wildlife Fund Book of Plants* (Garden City, N.Y.: Anchor Press/Doubleday, 1985).

2. Ibid.

Day 24. Fluids

1. Much of this chapter is drawn, with permission, from Bonnie Bainbridge Cohen's explorations with the fluid system. See "The Dynamics of Flow: The Fluid System of the Body," in *Sensing, Feeling and Action*, 66.

2. See Lorus Milne and Margery Milne, *Water and Life*, 6.

3. Kapit and Elson, *Anatomy Coloring Book*, 3.

4. Information on the lymph system from Shier, Butler, and Lewis, *Hole's Human Anatomy and Physiology*, 384–386. Fluids carry hormones from the endocrine system, neuropeptides as communication links for the nervous system, nutrients and waste materials in relation to the digestive system, lymphocytes for the defense system, and lactic acid from hardworking muscles, as well as lubricating, cushioning, and bathing tissues throughout our moving body.

5. Ibid.

6. Cohen, *Sensing, Feeling and Action*.

Day 25. Water

1. Sandra Postel, *Last Oasis: Facing Water Scarcity* (New York: W. W. Norton, 1992), 27. Research assistant Alexa Gilbert.

2. C. A. Hunt and R. M. Garrels, *Water: The Web of Life* (New York: W.W. Norton, 1972).

3. J. H. Brown and M. V. Lomolino *Biogeography*, 2nd ed. (Sunderland, Mass.: Sinauer Associates, 1998).

4. Ibid.

5. Ibid. Because the moon is closer to the earth, it exerts greater force; most shores experience two high and two low tides every twenty-four hours, with highs at dawn and dusk, lows at noon and midnight. They also exhibit two periods of extreme tides, and two low tides each month, corresponding to the new and full moons.

6. William P. Cunningham and Barbara Woodworth Saigo. *Environmental Science: A Global Concern* (Dubuque, Iowa: Wm. C. Brown, 1992), 306.

7. Hunt and Garrels, *Water: The Web of Life*.

8. See Lorus Milne and Margery Milne, *Water and Life* (New York: Athenium, 1964), and Alice Outwater, *Water: A Natural History* (New York: HarperCollins, 1966).

Day 27. Motion and Emotion

1. Daniel Goleman, *Emotional Intelligence*, 6–18, 289. "All [emotional] impulses are, in essence, impulses to act, the instant plans for handling life that evolution has instilled in us." LeDoux, *Emotional Brain*, 113. For information on the physiology of emotions and the limbic region, also see Candace Pert, *Molecules of Emotion: Why You Feel the Way You Feel* (New York: Simon and Schuster, 1999).

2. Davis, *Mapping the Mind*, 170–175.

3. Ibid., 173.

4. Ibid., 183–186.

5. Sewall, *Sight and Sensibility*, 69. Sewall writes, "In 1990, 650,000 new prescriptions of Prozac were written each month. In one year's time, this translates into 7.8 million new prescriptions . . . and Prozac is only one of a dozen or so such modern drugs in common use. The cascading use of antidepressants certainly suggests that depression is more than an individual affair."

Day 28. Sensuality and Sexuality

1. Natalie Angier, "Illuminating How Bodies Are Built for Sociability," *New York Times*, April 30, 1996, Science sec. Note: Dr. Sue Carter, of the University of Maryland in College Park, is renowned in the field of oxytocin research.

2. Ibid.

3. Olsen, *Bodystories*.

Day 29. Art and the Environment

1. Thomas Berry, "Art in the Ecozoic Era," *Art and Ecology* 51 (summer 1992): 42–48.

Day 30. Concerning the Spiritual

1. Philip Zaleski and Paul Kaufman, *Gifts of the Spirit: Living the Wisdom of the Great Religious Traditions* (San Francisco: HarperCollins, 1997), 112.

2. See Chungliang Al Huang, *Embrace Tiger, Return to Mountain: The Essence of Tai Ji* (Berkeley: Ten Speed Press, 1988). Huang is an author-philosopher and president of the Living Tao Foundation.

Day 31. Embracing Place

1. Pat Noll, personal correspondence. Also see Joanna Macy, "Working through Environmental Despair," in *Ecopsychology: Restoring the Earth, Healing the Mind*, 240.

The Boat, Washington Island, painting by Gail Olsen. Watercolor on paper.

SOURCES AND RESOURCES

The common instinct of language . . .
> Gallway Kinnell, "The Writer's Voice"

Each of us has authors whose work stimulates our imagination and inspires our intellect to further musing. The following books have remained within arm's reach during the writing of *Body and Earth*. For eloquence and specificity of language within the weave of nature and culture, I'm drawn to Barry Lopez's *Crossing Open Ground* and *About This Life* and John McPhee's *Annals of the Former World*. For sensuality and the capacity to unfold an irresistible tale engaging body and earth, Terry Tempest Williams's *Desert Quartet,* and David Abram's *Spell of the Sensuous* capture attention. A particular quality of gentleness, mixed with impeccable scholarship and dedication to home terrain, mark John Elder's *Reading the Mountains of Home* and Kathleen Norris's, *Dakota: A Spiritual Geography*. Musings on mind and body have been most satisfied by Joel Davis's *Mapping the Mind,* because of his explanation of the ways the environment and the arts shape the brain.

On the topic of body, *Bodystories: A Guide to Experiential Anatomy* serves as a companion guide to this text. Deane Juhan's *Job's Body: A Handbook for Bodywork* and Bonnie Bainbridge Cohen's *Sensing, Feeling, and Action* provide profound resources for body awareness. Consistently stimulating articles on the brain and gut by Sandra Blakeslee can be found in the science section of the *New York Times*. On the topic of mind, Dharma Singh Khalsa's *Brain Longevity* is helpful reading for anyone involved with Alzheimer's disease, and Frank R. Wilson's *The Hand: How Its Use Shapes the Brain, Language and Human Culture* reminds us of the historical importance of touch.

On the topic of earth, Lyall Watson's, *Dark Nature* and Richard Nelson's *Living with Deer in America* offer challenging perspectives. I was pleased to rediscover Adrienne Zihlman's *Human Evolution Coloring Book,* added to my collection in 1982 and resurfacing for this occasion. Several excellent field guides accompanied my sojourns, including two favorites, *Atmosphere* by Vincent Schaefer and John Day for Peterson Field Guides and *National Audubon Society's Field Guide to New England,* by Peter Alden and Brian Cassie. For specifics of geology, I turned to *Planet Earth* by Cesare Emiliani, and for body to Gerard Tortora and Sandra Reynolds's *Introduction to the Human Body: The Essentials of Anatomy and Physiology,* the way I turn to *The Chicago Manual of Style* to decide, finally, where to put a comma.

For creative writing itself, I have enjoyed, as have many others, the books of Natalie Goldberg, *Writing Down the Bones* and *Wild Mind,* along with her reflections on spirituality, art, and the environment in *Long Quiet Highway*. Contemplative practice techniques are well described in form and intention in Erich Schiffmann's *Yoga: The Spirit and Practice of Moving Into Stillness,* Al Huang's *Embrace Tiger, Return to Mountain; The Essence of T'ai Chi,* and *Authentic Movement: Essays by Mary Starks Whitehouse, Janet Adler and Joan Chodorow,* edited by Patrizia Pallaro. For those who teach and for those with a zest for learning, there are numerous texts available for investigation on the topic of body and earth. You will, no doubt, find your own favorites at this unique time in your life, addressing the particular bioregion where you live.

SELECTED BIBLIOGRAPHY

Abram, David. *The Spell of the Sensuous: Perception and Language in a More-Than-Human World*. New York: Random House, 1996.

Austin, James H. *Zen and the Brain: Toward an Understanding of Meditation and Consciousness*. Cambridge, Mass.: MIT Press, 1998.

Berry, Wendell. *The Unsettling of America*. San Francisco: Sierra Club, 1977.

Cohen, Bonnie Bainbridge. *Sensing, Feeling, and Action: The Experiential Anatomy of Body-Mind Centering*. Northampton, Mass.: Contact Editions, 1993.

Davis, Joel. *Mapping the Mind: The Secrets of the Human Brain and How It Works*. Secaucus, N.J.: Carol Publishing Group, 1997.

Devall, Bill, and George Sessions. *Deep Ecology: Living As If Nature Mattered*. Salt Lake City: Peregrine Smith Books, 1985.

Elder, John. *Reading the Mountains of Home*. Cambridge, Mass.: Harvard University Press, 1998.

Farhi, Donna. *The Breathing Book: Good Health and Vitality Through Essential Breath*. Markham, Ont.: Fitzhenry and Whiteside, 1989.

———. *Yoga Mind, Body and Spirit: A Return to Wholeness*. New York: Henry Holt, 2000.

Gershon, Michael D., M.D. *The Second Brain*. New York: Harper Collins, 1998.

Gorman, David. *The Body Moveable*. Guelph, Ont.: Ampersand Press, 1981.

Hawken, Paul. *The Ecology of Commerce, A Declaration of Sustainability*. New York: Harper Collins, 1993.

Hiss, Tony. *The Experience of Place*. New York: Knopf, 1990.

Hogan, Linda. *Dwellings, A Spiritual History of the Living World*. New York: W.W. Norton, 1995.

Huang, Chungliang Al. *Embrace Tiger, Return to Mountain: The Essence of Tai Ji*. Berkeley, Calif.: Ten Speed Press, 1988.

Hughes, J. Donald. *American Indian Ecology*. El Paso: Texas Western Press, 1983.

Huxley, Anthony. *Plant and Planet*. New York: Viking, 1990.

Johnson, Don Hanlon. *Body: Recovering our Sensual Wisdom*. Berkeley, Calif.: North Atlantic Press, 1983.

Juhan, Deane. *Job's Body: A Handbook for Bodywork*. Barrytown, N.Y.: Station Hill Press, 1987.

Jung, Carl G. *Man and His Symbols*. Garden City, N.Y.: Doubleday, 1964.

Kapit, Wynn and Lawrence Elson. *The Anatomy Coloring Book*. New York: Harper Collins, 1993.

Ledoux, Joseph. *The Emotional Brain*. Simon and Schuster, New York, 1995.

Levine, Stephen. *Who Dies? An Investigation of Conscious Living and Conscious Dying*. Doubleday, New York, 1982.

Leystron, Andrew, David Matless, and George Revill. *The Place of Music*. New York: Guilford Press, 1988.

Linklater, Kristin. *Freeing the Natural Voice*. New York: Drama Book Specialists, 1976.

Lopez, Barry. *About This Life: Journeys on the Threshold of Memory*. New York: Knopf, 1998.

———. *Arctic Dreams: Imagination and Desire in a Northern Landscape*. New York: Bantam Books, 1996.

———. *Crossing Open Ground*. New York: Random House, 1989.

Lyndon, Donlyn, and Charles W. Moore. *Chambers for a Memory Palace*. Cambridge, Mass.: MIT Press, 1994.

McPhee, John. *Annals of a Former World*. New York: Farrar, Straus, and Giroux, 1999.

McLuhan, T. C., ed. *Touch the Earth: A Self Portrait of Indian Exist*. Simon and Schuster, New York, 1971.

Nelson, Richard. *The Island Within*. New York: Vintage Books, 1996.

Norris, Kathleen. *Dakota: A Spiritual Geography*. Boston: Houghton Mifflin, 2001.

Olsen, Andrea. *Bodystories, A Guide to Experiential Anatomy*. Barrytown, N.Y.: Station Hill Press, 1991.

Orr, David. *Ecological Literacy, Education and the Transition to a Postmodern World*. Albany: State University of New York Press, 1992.

Pallaro, Patrizia, ed. *Authentic Movement: Essays by Mary Starks Whitehouse, Janet Adler, and Joan Chodorow*. London and Philadelphia: Jessica Kingsley, 1999.

Pearsall, Paul. *The Heart's Code*. New York: Broadway Books, 1998.

Pert, Candace. *Molecules of Emotion: Why You Feel the Way You Feel*. New York: Simon and Schuster, 1999.

Ramachandran, V. S., and Sandra Blakeslee. *Phantoms in the Brain: Probing the Mysteries of the Human Mind*. New York: William Morrow, 1998.

Roszak, Theodore, Mary E. Gomes, and Allen Kanner, eds. *Ecopsychology: Restoring the Earth, Healing the Mind*. San Francisco: Sierra Club Books, 1995.

Samuels, Mike, and Hal Zina Bennett. *Well Body, Well Earth*. San Francisco: Sierra Club Books, 1983.

Sauer, Peter, ed. *Finding Home: Writings on Nature and Culture*. *Orion Magazine*. Boston: Beacon Press, 1992.

Schiffman, Erich. *Yoga: The Spirit and Practice of Moving into Stillness*. New York: Simon and Schuster, 1996.

Sewall, Laura. *Sight and Sensibility: The Ecopsychology of Perception*. New York: Tarcher/Putnam, 1999.

Sheldrake, Rupert. *Dogs That Know When Their Owners Are Coming Home, and Other Unexplained Powers of Animals: An Investigation*. New York: Three Rivers Press, 2000.

———. *The Rebirth of Nature, The Greening of Science and God*. New York: Bantam Books, 1991.

Sylvia, Claire. *Change of Heart*. New York: Warner Books, 1998.

Tortora, Gerard J., and Sandra Reynolds. *Introduction to the Human Body: The Essentials of Anatomy and Physiology, Fifth Edition*. New York: Biological Sciences Textbooks/Wiley and Sons, 2001.

Tuan, Yi-Fu. *Topophilia: A Study of Environmental Perception, Attitudes, and Values*. New York: Columbia University Press, 1990.

———. *Space and Place, The Perspective of Experience*. Minneapolis: City University of Minnesota, 1977.

Watson, Lyall. *Dark Nature: A Natural History of Evil*. New York: Harper Trade, 1997.

Williams, Terry Tempest. *Desert Quartet*. New York: Pantheon Books, 1995.

———. *Refuge: An Unnatural History of Family and Place*. New York: Vintage Books, 1992.

Wilson, Frank R. *The Hand, How Its Use Shapes the Brain, Language and Human Culture*. New York: Pantheon, 1998.

Zajonc, Arthur. *Captured by Light*. New York: Bantam Books, 1993.

Zihlman, Adrienne. *Human Evolution Coloring Book*. San Francisco: Harper Perennial, 1982.

SUBJECT INDEX

Page numbers in **bold type** indicate primary discussion of topic.

SPECIAL THANKS TO:

Phil Pochoda, Ellen Wicklum, Mary Crittendon, Katherine Kimball, and the staff at UPNE for all they do to make a book whole.

Michael Burton, Karen Murley, and Jennifer Ponder for aspects of design.

Rosalyn Driscoll, Lisa Nelson, Harriet Brickman, and Susie Caldwell for editorial assistance.

Erik Borg for collaborating on photo projects.

All the artists and educators who contributed creative work and words.

Colleagues Penny Campbell, Rebecca Gould, John Huddleston, Andrea Lloyd, Tom Root, Peter Ryan, Peter Schmitz, and Stephen Trombulak for reviewing chapters.

Former students Jean Andersson, Matthew Brown, Ben Calvi, Greg Carolan, Alexa Gilbert, Eve-Lyn Hinckley, Corinna Luyken, Greg Horner, Eliza Ingle, Shruthi Mahalingaiah, Susan Mendez Prins, Charlotte Sikes, Andrea St. John, and many others for research, writing, and discussion.

Field Naturalists Laurie Sanders, Fred Morrison, and Jeff Meyers for outdoor adventures and expertise.

Bob Upham for contributing "local color" and global perspective.

Sansea and Dennis Sparling for sharing place.

My family—Steve, Josh, and Jonah —for supporting and sharing stories throughout.

ACLS (American Council of Learned Societies), the Nathan Cummings Foundation, and the Fetzer Institute for a 1998 Contemplative Practice Fellowship for writings on contemplative practice in education.

Middlebury College for Ada Howe Kent Faculty Fellowships, a Service Learning Grant, and an Environmental Studies Summer Research Fellowship for text-related projects.

Dancers, Sarah Garcia and Glenn Olds. Photograph by Bob Handelman.

ART INDEX

Anonymous multicultural images are indexed under country of origin.